THE HISTORY O]
AND
FIELD SYSTEMS

Books of similar interest published by Scottish Cultural Press:

SCOTLAND SINCE PREHISTORY: Natural change and human impact
edited by T C Smout

THE HISTORY OF SOILS
AND
FIELD SYSTEMS

edited by

S Foster

and

T C Smout

SCOTTISH CULTURAL PRESS

First published 1994

Scottish Cultural Press
PO Box 106
Aberdeen AB9 8ZE
Tel/Fax: 01224 583777

British Library Cataloguing in Publication Data
A catalogue record for this book is available
from the British Library

ISBN: 1 898218 13 7

Scottish Cultural Press acknowledges a subsidy from Historic Scotland
towards the publication of this volume

Printed and bound by
Bookcraft, Midsomer Norton, Bath, Avon

CONTENTS

FIGURES

PLATES

TABLES

CONTRIBUTORS

JOHN CATT, Rothamsted Experimental Station, Harpenden, Herts

DONALD DAVIDSON, Department of Environmental Science, University of Stirling

PIERS DIXON, The Royal Commission on the Ancient and Historical Monuments of Scotland, Edinburgh

ROBERT A DODGSHON, Institute of Earth Studies, University College, Aberystwyth

ALEXANDER FENTON, School of Scottish Studies, University of Edinburgh

SALLY FOSTER, Historic Scotland, Edinburgh

DAVID HALL, The Fenland Project, Department of Archaeology, University of Cambridge

RICHARD HINGLEY, Historic Scotland, Edinburgh

ROGER MERCER, The Royal Commission on the Ancient and Historical Monuments of Scotland, Edinburgh

JOHN MILES, Scottish Office Central Research Unit, Edinburgh

JOHN SHAW, National Museums of Scotland, Edinburgh

IAN A SIMPSON, Department of Environmental Science, University of Stirling

RICHARD TIPPING, Peebles, Scotland

DONALD WOODWARD, Department of Economic History, University of Hull

PREFACE

The Institute for Environmental History organised a seminar on the History of Soils at Battleby in December 1993, with support from Scottish Natural Heritage and Historic Scotland. It was obvious to those of us who had only a few weeks earlier attended a seminar on Medieval or Later Field Systems, organised by Historic Scotland in Edinburgh, that these two subjects were complementary. While some of the issues related to the preservation and management of medieval and later field systems are specific to this period, broader concerns, specifically the archaeological potential of soils have a far wider relevance. The decision has therefore been made to publish these seminars jointly, with financial assistance from Historic Scotland.

Soils have history – and a future. Ecologists, archaeologists, historians and soil scientists have in different ways increasingly realised the extent to which soil is not only capable of being changed by human action and by continuing natural forces, but has been so changed for millennia. In their interaction with the environment and the exploitation and harnessing of its natural resources, human beings have incessantly modified soils and defined the modern 'cultural landscape' in which soils are situated. None have had more impact than farmers, whose continuous and ingenious struggle to improve or maintain the fertility of the land has both created soils and altered their structure, with resultant changes in biodiversity, landscape character and profound social consequences.

Soils and associated landscapes of agricultural exploitation are therefore a repository of data about past human practice. Analysis of the structure, profile, biological and chemical components of soils offers unique insights into the past. They not only contain past artefacts and structures, but are artefacts in themselves. Conservation and management of these soils and landscapes is therefore essential for all those interested in the history, amenity, educational value and future of the countryside as a whole. Lessons about the long-term effects of soil erosion, decay and mismanagement can be learnt from the past, and we must now take these into account when considering the detrimental impact of global warming, forestry, erosion and atmospheric and other forms of pollution on this fragile resource. Soil (as the landscape as a whole) has a future, but what this is to be is very largely up to ourselves, just as in the past.

Sally Foster (Historic Scotland)
Chris Smout (Institute for Environmental History, St Andrews University)

Chapter 1

The Prehistory of Soil Erosion in the Northern and Eastern Cheviot Hills, Anglo-Scottish Borders.

Roger Mercer and Richard Tipping

Introduction

Archaeological surveys in the Scottish uplands have yielded valuable data on the extent, density, altitudinal range and broad chronology of successive settlement phases (Mercer 1991). In broadening the scope of archaeological enquiry from individual sites to entire landscapes, such field surveys have extended our understanding of prehistoric and historic population dynamics. Perhaps a more valuable achievement, however, has been the way in which such landscape-based archaeological data-collection has allowed the bridging of the gap between the social and natural sciences.

Techniques for understanding changes in the 'natural' landscape over the present interglacial, the Holocene (8300BC-Present), developed independently (Birks & Birks 1980; Lowe & Walker 1984), and are still commonly practiced without recourse to the record of human occupation. Increasingly, however, the influence of the landscape on people, and of people on the landscape, have come to be regarded as seminal problems in palaeoenvironmental research (e.g., Birks et. al. 1988; Chambers 1993).

Tensions arise from the application of different techniques to specific archaeological sites, because the spatial scales of enquiry most appropriate to some methods are not so to other approaches. These conflicts are difficult to resolve unless the scales of enquiry of different techniques are induced to overlap, and in this the adoption of macro-landscape survey methods in archaeology is one of the most significant advances. Archaeological investigation of an entire river catchment, for instance, has a resonance for the geomorphologist, who has long understood that the drainage basin is 'the fundamental geomorphic unit' (Chorley 1969, 30). This scale is also most appropriate for the palynologist, who has difficulties resolving structure and change in vegetation at less than several tens of metres around a pollen site (Jacobson & Bradshaw 1981). The adoption of a suite of techniques, archaeological and palaeoenvironmental, suited to one unifying spatial unit, has the potential to move beyond providing a 'background' to human settlement, and to explore the complex 'feedback' linkages between anthropogenic and natural processes, and to generate a more holistic understanding of Holocene landscape evolution.

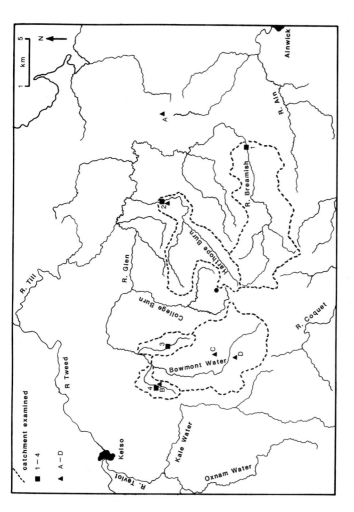

1.1: The drainage system of the Cheviot Hills in south east Scotland and north east England, depicting as dashed outlines the drainage basins investigated. The Cheviot (815m OD) is indicated by the black circle between the Harthope and College Burns. The archaeological survey area within the Bowmont Valley (Bowmont Water) is also outlined. The localities of reaches within the investigated drainage basins are depicted, as are those of radiocarbon-dated pollen sites within the Cheviot Hills.

Key: ■ *the river reaches dated by radiocarbon* : ▲ *radiocarbon-dated pollen sites relevant to the study*
1 - River Breamish at Powburn; **2** - Wooler Water at Wooler; **3** - Halter Burn; **4** - Yetholm Loch. **A** - Camp Hill Moss (Davies and Turner 1979, *New Phytologist*); **B** - Yetholm Loch; **C** - Sourhope; **D** - Swindon Hill; **E** - Mow Law 'A' and 'B'; **F** - Quarry Knowe.

The Bowmont Valley

In response to the possibility of widespread afforestation in this upland catchment, draining the northerly slopes of the Cheviot Hills in southeast Scotland, south east of the market town of Kelso (Fig 1.1), Historic Scotland funded a comprehensive five-year archaeological field survey in what is one of the most spectacular archaeological landscapes in the country (RCAHMS 1956).

The survey area covers the whole of the upper Bowmont Water drainage network from just south of Town Yetholm to the drainage divide north and west of The Cheviot itself (Fig 1.1). The total relative relief is 623m, the basin area is *c*52.55km², and basin length is 11.75km. Auchope Cairn (737m OD) is substantially higher than other parts of the watershed, where summits rise only tens of metres above a high-level summit surface of 490-520m OD. Valley-side slopes are commonly convexo-concave. Slope classes (Young 1972) are:

level to very gentle	0.0 - 3.0°	5.6% of the study area
gentle	3.1 - 8.0°	14.1%
moderate	8.1 - 14.5°	34.5%
moderately steep	14.6 - 18.0°	18.5%
steep	18.1 - 31.0°	24.9%
very steep	31.1 - 42.0°	2.3%

The smoothness, and the dominant gentle to moderate slopes, are the products both of a uniform geology of Lower Devonian andesite lavas, and of the periglacial modification of till and bedrock slopes (Harrison 1993).

Despite the uniformity of the solid geology, there is considerable complexity in the types and distribution of soils. The five soil series (Muir 1956) vary primarily in drainage state. Above the alluvium of the Bowmont Water and its tributaries, poorly drained non-calcareous gleys of the Atton Series are found on the lowermost slopes (gentle to moderate slope angles), on clayey till and solifluction sheets. Above these, on moderate to steep slopes, are found Sourhope series soils, freely drained brown forest soils. Freely-drained peaty podzols of the Cowie Series occur on the extensive, gentle plateau surfaces at intermediate altitudes (350-450m OD). On very steep slopes, above and below these plateaux, skeletal soil complexes are very common.

The good correspondence between, in particular, skeletal soils and steep to very steep slopes, implies an element of current soil instability on slopes steeper than *c*20°, around 28 percent of the valley. However, active erosional forms in the valley are very few. In the Pennines, slope angles >36° generally do not allow continuous vegetation cover (Young 1972). In the Bowmont Valley slopes steeper than 31° comprise <3 percent of the area, and this is one reason for the virtually complete ground cover of thick and matted Cheviot turf at the present day. By far the most extensive vegetation is grassland, ranging from base-rich *Agrostis-Festuca* communities to acid and waterlogged *Nardus* stands (King 1962; King & Nicholson 1964). *Calluna* (ling heather) heath is common on less intensively grazed areas (King 1960; Hunter 1962).

The Bowmont Valley Archaeological Survey

The work was undertaken by students of the Department of Archaeology, Edinburgh University, with support from a team of professional archaeologists, under the direction and leadership of Roger Mercer.

The survey was based upon two primary means of data collection; intensive field walking and the inspection of air photographic cover in the valley. The latter was founded upon the possession and use of total vertical cover of the whole valley taken in low sunlight on a summer morning in 1947 and enlarged to a scale of 1:2500. Elements of other aerial coverage of the valley were used, as was available, in the form of oblique photography both pertaining to the RCAHMS air photographic programme, and sorties undertaken by Professor D W Harding, University of Edinburgh. All unitary sites (house sites, enclosures and structural complexes) were recorded in detail at scales ranging from 1:50 to 1:500. The whole valley with every recorded trace of agricultural organisation, (cairn fields, cord rig, terraced agriculture, rig and furrow, field enclosures) was recorded at 1:1000 scale with detail, eradicated since 1947, but visible on aerial photographs of that date, included. The total of unitary sites in addition to identifiable separate components of agricultural activity comprises well over 1,300 (see Mercer et.al. *forthcoming*).

The identification of this intense distribution of traces of human activity, and the intricate detail wherein it has proved possible to record it, depend very largely upon the unique quality of the Cheviot turf. Slow-growing, intensively cropped by sheep and, over large tracts, undisturbed by cultivation for at least a millennium and a half, it has allowed the survival, as visible field evidence, of traces of structures that were, when built, substantially constructed of timber. The timbers have, of course, long ago disappeared, but the sockets and slots into which they were originally set survive to this day as depressions and grooves still visible to the trained eye. Earthworks survive, generally, in a relatively undenuded state, due to the lack of subsequent disturbance referred to above, while stone structures are also present, with good building stone available either from field clearance or quarrying.

To the modern observer two aspects of this field surveying environment call immediately for comment. First there is the ample evidence for timber construction, especially in the earlier witnessed phases of prehistoric development (at a stage before and around the middle of the first millennium BC), in circumstances that, today, are very largely treeless, but must, then, have supported extensive, and carefully managed, stands of, probably coppiced, timber. Secondly there is the very intensity of the field remains already referred to, which from this date characterise the valley, and which must suggest substantial human population in an area now occupied by relatively few people. A third question which must immediately press itself upon the observer is whether the apparently sudden first appearance of widespread human activity in the valley, at a stage conventionally dated by archaeologists to the earlier centuries of the first millennium BC, is a reflection of reality or not. This is a vital question upon which Tipping's investigations have thrown much light and which suggest that, such is the vigour of this episode, that it has eradicated or buried earlier traces of activity that reach back at least another millennium, but which are now only detectable by us through the medium of palaeofluvial and palaeobotanical studies.

The archaeological survey revealed an extraordinarily high density of remains, although the utmost caution must be used in using field remains of structural or agricultural practices (particularly the latter) as diagnostic chronological indicators. The earliest archaeologically recognisable presence of settlement in the valley may be represented by the high altitude hill-top enclosure on Hownam Law – an enclosure of some 10 hectares with, set within its *enceinte*, over 100 small platforms upon which houses would have been constructed. The evidence of early date for this site, set where the Bowmont Water debouches onto the Merse, on the outermost of the high spurs of the Cheviots, is entirely on comparative grounds. This is so both specifically, with the excavated evidence for similar platforms set within a high altitude enclosure at Eildon Hill North, near Melrose (Owen 1992) which produced C14 dates relating to the primary occupation ranging between 1070-650 cal. BC, and in more general terms. There is now a broadly dispersed series of high altitude enclosures in Scotland (see Ralston & Smith 1983; Sanderson et.al. 1988; Mercer 1991) which a conjunction of different strands of evidence might suggest consistently relate to this, conventionally, Middle Bronze Age date.

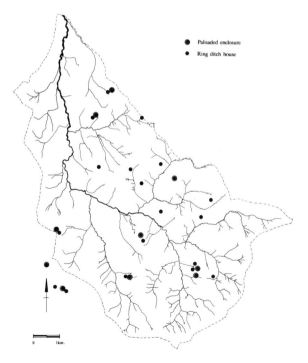

1.2 : Distribution of RDHs and palisaded enclosures in the Upper Bowmont Valley.

The next chronological 'grab-rail' that archaeologists have precariously erected would appear to relate to a horizon (?) of activity where large (10m diameter plus) round houses appear in Scotland (Mercer 1985, 59-95; Hill 1982a) (Fig 1.2). The circumstances of the Bowmont Valley, and of large tracts of the lowland zone, allowed these houses to be built of timber with a conical thatched roof, upon vertical supports, with an external screen-wall composed of closely juxtaposed vertical timbers set within a penannular foundation-groove. Such 'ring-groove' houses (RGHs), named for their principal surviving field trace, are generally encountered in the Bowmont Valley enclosed within what were fenced enclosures, of up to 0.6ha in extent, containing up to twelve houses in an apparently carefully arranged pattern set beside a focal 'street', with their doors generally facing away from this thoroughfare. Such houses have not been adequately and consistently dated in Lowland Scotland, but their stratigraphical position *vis a vis* other dated deposits on sites in East Lothian (Hill 1982b; Triscott 1982) suggest a date prior to 500 cal. BC, while the location of an iron spearhead at Hayhope Knowe in the Bowmont Valley during Mrs C M Piggott's excavations (Piggott 1949) indicates that, in this instance at least, the occupants were iron-using and therefore, perhaps, unlikely to predate 700 cal. BC. Hill (1982a) has made the astute observation that while the houses (the *sine quae non* of settlement) may well be chronologically diagnostic, the form of enclosure of the settlement will respond far more flexibly to local social, economic and environmental circumstances, and it is indeed true that radiocarbon determinations for palisaded enclosures embrace the whole latter half of the first millennium cal. BC. Nevertheless the earliest dates for a range of such sites extending through both English and Scottish border country fall in the sixth–seventh century cal. BC (Burgess 1984). Such settlements (e.g. Hayhope Knowe) are frequently associated with a type of cultivation which, it is now almost certain, persisted over a very long period, of spade-dug trenches set within 1m-1.5m of each other with the dug material piled on the intervening berm (see Topping 1989), and known as cord-rig. The rigs are almost always aligned across the contour, i.e. up and down slope, and were presumably created with the two-fold purpose of *draining* the soil and *elevating* the tilth to catch the maximum strike of the sun.

The population of the Bowmont Valley at this time, by the middle of the first millennium BC, is hard to estimate, of course, but at least eight such settlements are known in the valley (not necessarily all contemporary) with at least six of these great houses in each, which perhaps allows us to suggest a population of 500 plus.

Within the valley itself, and elsewhere, there does appear to be a consistent stratigraphical relationship between RGHs and a developed house-type, known again by its surviving features, as the ring ditch house (RDH). Such houses would appear to form an evolutionary improvement in design upon the RGH in that they comprise an artificially lowered area, probably for the in-housing or wintering of milch cattle, 'wrapped around' the RGH concept. As well as further reinforcing the stability of the conical house design, this development would have furnished a *cordon* (if hardly *sanitaire*) of insulation to the central dwelling component. The artificially lowered 'ring ditch' element would have contained the dung of the sheltered animals, preventing this spewing forth on to the house floor - to be mucked out every spring for deposition on the fields. In no instance in the Bowmont Valley can it be demonstrated that an RDH is a primary component of a curvilinear palisaded settlement, while there are at least three instances where the insertion of RDHs on these sites has over-ridden what is clearly, by this point, a defunct palisade trench. (This observation sets aside the unique rectilinear palisade containing two RDHs at Greenborough). Radiocarbon evidence for the date of RDHs in Lowland Scotland (see Hill 1982a) is of relatively high quality and, so far, shows remarkable uniformity. A series of dates from Dryburn Bridge, East Lothian (Triscott 1982) and Broxmouth, East Lothian (Hill 1982b) show, crudely, an overlap with RGH construction (they may well be, to some extent, contemporary structure-types of different function) but with dates reaching up into the fourth century cal. BC, later than the admittedly poorly dated RGHs of equivalent size. In the Bowmont Valley there does appear to be a stratigraphical relationship between the two whereby the RGH is consistently the earlier, as well as apparently contemporary with enclosing palisades. RDHs oversail redundant palisades and are associated (with the possible exception of Hownam Law, see above) with the earliest earthwork fortifications in the valley – enclosures that we would conventionally term hill-forts (Fig 1.3).

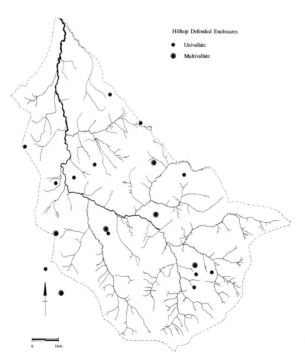

1.3 : Distribution of hilltop defended enclosures in the Upper Bowmont Valley.

The hill-forts of the valley can be divided, broadly, into three categories. We take from Macinnes (1984) her categories 'plateau fort', and 'promontory fort' which have an altogether wider application in the Scottish borderlands. Of her 'citadel forts' there appear to be none in the upper Bowmont Valley (with the possible exception of Hownam Law). The Bowmont also produces what the writer has chosen to call Knoll-forts – small (0.5ha), often multivallate, enclosures set on prominent knolls or hill tops (like those at Camp Tops, Kip Knowe and Sundhope Kipp).

This latter class of monument may well be complex in development and there is a consistent relationship in, at least, the earlier stages of development with RDHs. Indeed RDHs are present on all hill-fort sites in the valley with the exception of 'The Castles', Blackborough, Kipknowe and the later, earthwork fortified, phase of Hayhope Knowe. In the latter two of these instances the earthwork would appear to be manifestly unfinished.

We might therefore choose to infer that vallate hill-fort construction (with the possible exception of Hownam Law) began during a phase when RDHs were under construction and, indeed, where RGHs appear within the *enceinte* there is evidence in two instances that they may pertain to earlier, now eradicated, palisaded enclosures.

Any remotely reliable understanding of hill-fort development within the valley must depend upon a major programme of excavation which is not currently envisaged. Sites like 'The Castles' and Park Law would seem to be the subject of complex multiphase development. This would appear to have come to, at least, a temporary end, when farmsteads which bear morphological resemblance to examples which, when excavated, yield evidence of Roman influence. Hill (1982a), feeling, appositely, that the term Romano-British was unsuitable, has termed these farmsteads 'Votadinian'.

Hill's (1982a) examination of the available radiocarbon dating evidence from settlements (often called farmsteads and scooped settlements) of this type, that have undergone excavation, has shown a consistent tendency for a prolonged developmental chronology which in one or two instances commences with the existence of RGHs (see above). Routinely, however, radiocarbon evidence indicates an origin for these sites in the first and second centuries cal. BC (see Hill 1982a). They consist of enclosed farmsteads, with two-to-five houses of relatively small size set within them, with a propensity for those houses to be built upon a stone foundation the later in time that their construction takes place - a development that may reflect upon changing environment and the increasing scarcity of timber. Radiocarbon evidence from excavated sites in Northumberland (Kennel Hall Knowe and Belling Law) would suggest that construction of those sites persisted well into the third century cal. AD (Fig 1.4).

Over 40 such settlements are known in the upper Bowmont Valley, and if an extended family of say, 20, are to be associated with each site a population of the order of magnitude of 800 for the area (16 per sq.km) would perhaps be acceptable. Such a population might require 1000ha for its subsistence arable base (see Mercer 1981, 231 *et.seq.*), a figure which matches well with the extent of terraced agriculture in the valley which, it is argued (on, admittedly, flimsy grounds) is almost certainly the agricultural strategy of this period (see below).

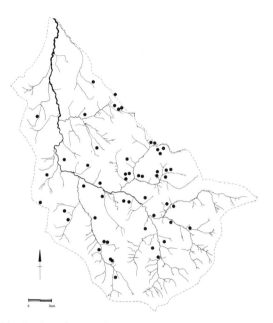

1.4: Distribution of scooped settlements and farmstead enclosures
of 'Votadinian' type in the Upper Bowmont Valley.

Two other features relating to the early occupation of the valley need to be briefly considered. The first is the existence within the valley of unenclosed platform settlements (Fig 1.5). These comprise a series of platforms dug back into the hillside with the resultant spoil cast forward to form an 'apron' which in combination with the dug level surface forms a level platform upon which to build a house. It is easier to walk between such houses if they are all at the same level so, naturally, there is a tendency for these platforms to 'follow the contour'. The unenclosed settlements, therefore, usually comprise a string of four or five such platforms running along the side of a hill slope, often further distinguished to the modern eye by a differential vegetation cover. Now it will be immediately clear that we are not dealing here with a cultural artefact, a diagnostic indicator of differential human design through time, like the RGHs(?), RDHs and 'Votadinian' settlements. Platform building is a *technique* of producing a house-stance which has been reinvented on many occasions in many different locations. Consequently the induction of radiocarbon dating evidence from excavated examples in the Borders area is unlikely to assist with the chronological assignation of these features – as indeed it does not. Houses built upon platforms, whether enclosed or not, have produced dating evidence from the mid-second millennium cal. BC to the Medieval period (see Owen 1992; Jobey 1980; Stevenson 1941). One issue may, however, assist us (and may not as we shall see). The *size* of the platform must reflect the size of the house built upon it and in all instances the platforms in Bowmont Valley are relatively diminutive. In no instance does the 'building space' exceed 8m diameter (indeed on Hownam Law they are even smaller) and, therefore, a house of standard RGH or RDH type cannot be built upon them. We may have to regard platform settlements as an intermittent 'background noise' to settlements of more culturally diagnostic type throughout the whole period of prehistoric (and later?) occupation of the valley.

The second feature which, perhaps, is worthy of especial mention is to be associated with the hill forts at Craik Moor and Park Law. At these two sites, at what is clearly the latest and, apparently, terminal stage of the hill-forts' development, two great vertically faced walls, up to 4m in thickness, survive almost intact, cutting off the promontories from easy approach in both cases. In both instances the defence is penetrated by a simple gateway 2-3m wide, which in the case of Craik Moor has been blocked. At Park Law it is unlikely that this rampart could be earlier than the latest prehistoric activity on the site. The writer can only suggest, there is very little evidence, that this activity may relate to very early medieval rather than prehistoric date.

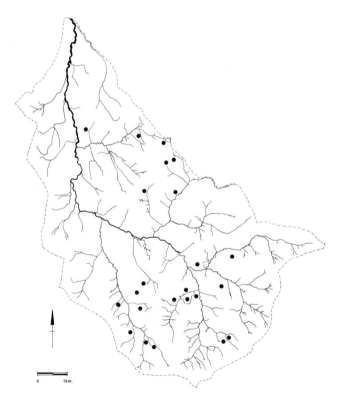

1.5: Distribution of unenclosed platform settlements in the Upper Bowmont Valley.

Palaeoenvironmental investigations

At an early stage, Richard Tipping was invited to devise a programme of research designed to reconstruct key elements of the Holocene landscape. The principal technique employed was palynology, using a network of six pollen sites within and immediately outside the archaeological survey area (Fig 1.1), correlations between sites obtained through a comprehensive radiometric dating programme. From these sites was constructed a very detailed understanding of the temporal and spatial patterning exhibited by the natural woodland, and the land-use changes that led to the woodland's disappearance.

One critical issue in the survey was to define how intensive the periods of densest human occupation were. Since one cannot assume contemporaneity of individual settlements in one morphological type without extensive and expensive excavation and intensive dating (and not necessarily even then), the easiest way to monitor population density is probably through the palaeoenvironmental record. However, although major phases of anthropogenic impact can be distinguished from more limited incursions, there is as yet no convincing way to determine from palynology the absolute scale or intensity of anthropogenic activities.

One other approach to this problem is to attempt to identify periods when human activities destabilized the landscape sufficiently to induce soil erosion on scales greater than would normally be expected. There are a number of assumptions in this aproach, not least the suggestion that soil erosion is unambiguously associated only with intensity of settlement. This is incorrect. As will become clear, the unravelling of causal agency in the instigation of soil loss is a complex procedure, and conclusions often equivocal.

Establishing a soil erosion chronology for the Cheviot Hills

Soils themselves rarely contain a record of soil losses. The truncation of soil horizons through erosion can occasionally be identified, but the dating of such events is very rarely able to be understood. Sediments move downhill, and only rarely are long-term records preserved on slopes, such as in talus accumulations (van Mourik 1986; Innes 1983).

In establishing a chronology of soil erosion, there is a need to analyze sediment sinks, areas of long-term sediment storage (Brown 1987a). Of these, two have been explored most fully, river terraces (Gregory 1983; Gregory, Lewin & Thornes 1987; Starkel, Gregory & Thornes 1991; Needham & Macklin 1992), and alluvial fan deposits (Brasier, Ballantyne & Whittington 1986; Brasier & Ballantyne 1989; Harvey et. al. 1981; Harvey & Renwick 1987; Tipping & Halliday 1994). Both offer the potential to provide comparatively long-term records of sediment accumulation, and by the burial and incorporation of *in situ* organic deposits, afford the opportunity of radiocarbon dating. Alluvial fans suffer the disadvantage of deriving from very localized sediment sources, and erosion chronologies derived from these need not relate to major changes occurring within the catchment. Equally, however, unless the sources of sediment within riverine deposits can be traced (e.g., Macklin, Passmore & Rumsby 1992), river terrace deposits provide only generalized data on sediment movement.

It is also true that such sediment sinks are separated in space from the most probable sources of sediment, the soils themselves. It needs to be assumed that soil erosion is directly causal in the deposition on valley floors of fluvial sediments. This is not always the case, in that stream-bank erosion can result in the transfer of sediment from older river terraces without contributions from eroding soils, but for substantial terrace accumulations the linkage between slopes and rivers can be accepted (Lewin 1981).

It has been suggested that valley floor deposits can be divorced in time from the erosion that provided the sediment (Burrin 1985), through lags in the processes of sediment transfer leading to short-term but significant delays in sediment deposition on valley floors, but this has been challenged by Richards et. al. (1987), because such lags would be unlikely to be so lengthy as to be measurable through, for instance, radiocarbon dating.

Although valley floors and terraces provide long-term sediment sinks, these are not permanent, and in upland environments such as the Bowmont Valley, characterized by very steep river gradients, gravel-dominated fluvial sediments and highly active channels (cf. Newson 1981), erosion of fluvial deposits is common. Thus, the Bowmont Water itself contains a record of past fluvial activity extending in time from the present only some 300-400 years (Tipping *in press* a). This necessitated the examination of other valleys draining the Cheviot. Four separate reaches were examined within the valleys of (i) the River Breamish, (ii) the Wooler Water, (iii) the Halter Burn and (iv) a small unnamed stream flowing into Yetholm Loch (Figure 1.1).

A soil erosion chronology for the Cheviot Hills

The methods, specific localities, details of the sediment types forming the various terrace accumulations, and the deposits radiocarbon dated at the NERC Dating Laboratory, have been described in Tipping (1992) (see also Tipping *in press* b). These data are not reproduced here. The relevance of each radiocarbon date (Table 1.1) is briefly described here. Dates are presented as either calibrated ages (cal. BC or cal. AD) or radiocarbon ages (BC or AD). Calibration is by the University of Washington CALIB 2.0 program (Pearson & Stuiver 1986; Stuiver & Pearson 1986).

(i) The Halter Burn.

The oldest Holocene terrace accumulation in this small (10.5km²) catchment, a tributary of the lower Bowmont Water (Fig 1.1), has a complex stratigraphy of channel gravels and fine-grained overbank sediments. At the base of the fill are the roots of at least one shrub or young tree (unidentified), rooted into a partly eroded gleyed soil. It is thought that the initial deposition of gravels eroded the soils lining the valley floor, and quite probably killed the shrub or tree. The radiocarbon date (SRR-3664; Table 1.1) of 2557-2407 cal. BC should date closely the earliest Holocene aggradation in this valley.

LAB.. NO	FRACTION DATED	^{14}C AGE BP & 1σ	$\delta^{13}C$ (%)	CAL. AGE RANGE BC/AD AT 1σ
(a) HALTER BURN				
SRR-3664	whole sample	3940 ± 45	-30.5	2557 - 2407 BC.
(b) WOOLER WATER				
SRR-3658	wood	3695 ± 45	-28.7	2184 - 2034 BC
(c) POWBURN QUARRY				
SRR-3659 a	'humic' carbon	1890 ± 45	-29.5	68 - 138 AD
b	'humin'	1850 ± 45	-29.2	89 - 225 AD
	average	1870 ± 32		85 - 142 AD
SRR-3660 a	'fine' fraction	2565 ± 45	-29.6	803 - 769 BC
b	'coarse'	2390 ± 40	-30.1	517 - 400 BC
SRR-3661 a	'coarse' rootlets	1800 ± 50	-27.0	133 - 316 AD
b	'humic' carbon	2275 ± 40	-29.3	395 - 267 BC
c	'humin'	1995 ± 50	-29.3	89 BC - 62 AD
SRR-3662	whole sample	10025 ± 55	-30.4	n.a.

n.a: not applicable

Table 1.1 : Details of radiocarbon dates pertaining to the erosion chronology

(ii) The Wooler Water at Wooler.

Clapperton, Durno & Squires (1971) reported a terrace fill, principally comprising coarse gravels and cobbles, between 2.0 to 2.5m thick, overlying a laterally extensive peat bed. Pollen analyses suggested a late Holocene age, post-3000 BC.

This terrace is named (Tipping *in press* b) the Earle Mill Terrace, and is the oldest Holocene terrace mapped in the valley. A second, lower and later, but undated, terrace fill is also mapped. A radiocarbon sample on wood fragments from within the topmost 10cm of peat, immediately beneath coarse sands, yielded an age (Table 1.1) of 2184-2034 cal. BC, suggesting that deposition of the fluvial sediments post-dates c2100 cal. BC.

(iii) The River Breamish at Powburn.

West of the Newcastle-Wooler road (A697), and east of the Cheviot uplands at Ingram, the River Breamish has one major terrace fill. Up to 6m of coarse gravels and cobbles, probably deposited as braided river deposits, have infilled the valley, producing a terrace surface 5.5km long and up to 1.5km wide. Younger terraces are found downstream of the A697.

Four radiocarbon samples were obtained. Date SRR-3660 (Table 1.1) is from a thin peat at or close to the basse of the gravels. Although the two dated organic fractions agree at 2 σ, they define the commencement of gravels only to a period between 820 to 390 cal. BC. Also close to the base of the gravels, but at the edge of the valley floor and stratigraphically higher, and only buried after the thalweg of the valley had been infilled with approximately 2-3 m of gravels, peat dated by SRR-3661 is interpreted to indicate continued gravel deposition after 133-316 cal. AD (Table 1.1). The three organic fractions differ at 1 σ, but differences between the ages of the 'coarse' rootlet and humin fractions are resolved at 2 σ. Rootlets can in many instances be suspected of intrusion from overlying sediments, but in this case the 2m of overlying gravel effectively nullifies this error, and this component is regarded as the one most likely to be *in situ*, and to reflect the age of the peat.

Assay SRR-3659 shows excellent agreement between fractions, and dates to between 85 and 150 cal. AD (Table 1.1) the infilling of one abandoned braided river channel with peats and organic-rich clays. Gravel continued to be deposited after this, as the channel fill lies 2.55 m below the terrace surface. A similar but much less organic channel fill higher within the gravel sequence yielded a radiocarbon age of *c*8000 BC (SRR-3662; Table 1.1), and does not contribute to an understanding of the fluvial sequence.

Archaeological evidence (Welfare 1992) suggests that terrace construction had certainly ceased by the Medieval period, when rig-&-furrow cultivation developed on the terrace surface.

(iv) Yetholm Loch.
An unnamed stream between the Kale and Bowmont Waters (Fig 1.1) drains a subglacial meltwater channel, and flows into Yetholm Loch. Within this small loch, a continuous sediment stratigraphy dating from the earliest Holocene has accumulated (Tipping *in press* b): lakes are more efficient sediment sinks even than river valleys. Within this sequence one extraordinary band of coarse silty sand, 8cm thick, disrupts the otherwise uniform clays and organic muds. The base of this band could not be radiocarbon dated, but has an interpolated age of *c*200 cal. BC. Organic-rich muds formed immediately following cessation of this minerogenic inwashing have a radiocarbon date of 204-346 cal. AD, indicating an end to this phase of soil erosion and transfer to the lake within the third or fourth centuries AD.

(v) Summary.
From these data, two phases of major landscape instability in the prehistoric can be defined for the northern and eastern Cheviot Hills. The earliest occurred in the later third millennium cal. BC, the Early Bronze Age, and is recorded in both the Halter Burn and Wooler Water catchments. The date of cessation of this phase cannot be established. Both these valleys contain later terrace fills, but these cannot be dated, and it is not known whether they relate to the second prehistoric episode of accelerated fluvial activity, commencing at or before 400-

200 cal. BC, the Late Iron Age, synchronous in both the Breamish and Yetholm Loch valleys. This phase ended at Yetholm Loch between 200 and 350 cal. AD.

Discussion

Prehistoric soil erosion in the Cheviot Hills was periodic. Episodes of soil stability were interrupted by seemingly shorter phases where soils and sediments were transferred from hillslopes to the valley floors of several, and probably all, catchments draining The Cheviot.

By themselves, these geomorphic records do not provide an indication as to the causes of soil erosion, only a chronology. To establish cause requires the examination of parallel proxy records, but a cause cannot be demonstrated, merely suggested as being most probable. Two primary causes, anthropogenic activity and climate change, have been most frequently postulated in the formation of river terrace fills, and their relative significance remains unclear (Macklin & Needham 1992). In discussion of the two episodes of prehistoric soil instability in the Cheviot Hills, analysis is better served than in many regions in that archaeological and palynological records pertaining closely to the geomorphic data are comprehensive, due principally to the Bowmont Valley Survey itself. Nevertheless, one additional assumption is necessary, that these lines of evidence pertain to a wider region than the one Cheviot valley from which they were obtained, and that they are typical of the Cheviot Hills.

(i) Late Iron Age soil instability.

This phase is discussed first because the interpretative principles are more easily explored, and the lines of evidence are in many respects clearer.

Comparison of different proxy records is made easier when they are derived from the same source. In the paleoenvironmental reconstructions within the Bowmont Valley, the lacustrine record at Yetholm Loch has exceptional value. Although the amount of mineral matter restricts the opportunity to radiocarbon-date closely the onset of minerogenic inwashing, it is possible to match the sediment- and pollen-stratigraphic records more closely than at sites where correlation by chrono-stratigraphic means is the only way of establishing links.

Interpretations of the records are complex (Tipping 1992), but associated with the phase of soil inwashing is the most extensive and complete woodland clearance recorded in the pollen diagram (Tipping *in press* b). Clearance was of a totally different character to earlier prehistoric events, with the removal of virtually all woodland in the pollen catchment, seemingly very rapidly. Farmers grew oats and rye, pasture was maintained, and hay meadows seen today as traditional (Hughes & Huntley 1988) may have been established. This represents the establishment of an organized and planned landscape, not the pattern of small-scale assarts prior to the Late Iron Age. Significantly, at two more of the very few lacustrine sites palynologically studied in northern England and southern Scotland, wholesale deforestation is associated with minerogenic inwashing or the reworking of old organic matter, at Thorpe Bulmer and Neasham Fen, on the Tees coastal plain (Bartley et. al. 1976).

A comparable pattern of near-total woodland clearance is recorded at Sourhope (Fig 1.1), deep within the Bowmont Valley, slightly later in age, at *c*50 cal. BC. This pattern of major clearance is identified throughout southern Scotland and northern England (Dickson 1992; Dumayne 1992, 1993; Fenton-Thomas 1992; Tipping *in press* b, d; Turner 1979; Wilson 1981; van der Veen 1992), and although diachronous, is clearly a predominantly Late Iron Age phenomenon.

This extraordinary intensification of farming practice should find corollaries with the archaeological record. One such link is with the construction of hill-forts (Macinnes 1982; Burgess 1984; Jobey 1971; Smith 1988-89; Rideout & Owen 1992), loosely dated to the mid-first millennium cal. BC, but perhaps a 'better fit' is with the widespread distribution of 'enclosed farmsteads' and 'scooped settlements' in the Bowmont Valley and elsewhere in southern Scotland and northern England. Currently available radiocarbon dates from broadly parallel excavated sites elsewhere would suggest this intensification of settlement to have taken place during and after the mid-third century cal. BC (see above). It was presumably the advanced state of native farming and its high productivity that allowed Roman military strategists the option of permanently stationing a massive garrison in the area, with all its consequent heavy engineering demands.

A correlation of the geomorphological evidence with the anthropogenic record, both paleoecological and archaeological, is clear. This is strengthened by the absence of evidence in the late prehistoric period for major changes in climate which could have served as a triggering mechanism (Lamb 1981). Accelerated fluvial activity is common in the British Isles in the Iron Age (e.g., Brown 1987b; Brown & Barber 1985; Brown & Keough 1992; Harvey & Renwick 1987; Passmore *et. al.* 1992; Robinson & Lambrick 1984; Shotton 1978), but synchroneity is not apparent, and human agency is near-universally argued to have been causal.

It is possible to see in the archaeological record of the Bowmont Valley a response to the soil erosion which affected hillslopes. In every instance the scooped settlement/enclosed farmstead development, here dated from the mid-third century cal. BC on the basis of extrapolated radiocarbon dates, is intimately associated with proximate blocks of narrow terracing, which would appear to be an attempt to conserve hillside tilth from erosion. This ubiquitous cultivation technique is by far the most widespread of any subsequent period in the valley and reached altitudes in excess of 350m OD (Fig 1.6). Generally speaking, the rare patches of cord rig that survive are situated scattered around the upper limit of the terracing, which suggests that substantial tracts of it were destroyed by the new regime. This deliberate, and highly labour-intensive, reversal of previous arable strategy may suggest a change of 'policy' along the lines indicated. On gentler slopes, still running along the contour, a narrow rig, sometimes interspersed with apparently stressed furrows, would appear to respect the terracing at nearly all points and would appear to be a cognate equivalent. This complex of agricultural techniques had been long assumed to be of Anglian or Medieval date largely, apparently, on the grounds that such widespread agricultural activity could hardly be prehistoric. RCAHMS investigators (1956), working in

Roxburghshire, clearly had misgivings about this, but it was not until Halliday's and Topping's work in the late 1970s (Halliday 1982 and Topping 1983) that the possibility of a prehistoric date was firmly raised.

There are, of course, many reasons for the development of cultivation terracing, but one likely explanation is as a soil conservation measure.

What is also apparent, however, from the Yetholm Loch stratigraphy, is that the apparently severe soil losses had no debilitating long-term effect on agricultural productivity. Cleared woodland gave way to cultivated ground with no apparent lag. Much more needs to be understood of the linkages between the archaeological, palynological and geomorphic records, but it can be suggested that while soil conservation was of obvious concern, soil erosion had not reached the point at which agricultural output was being adversely affected.

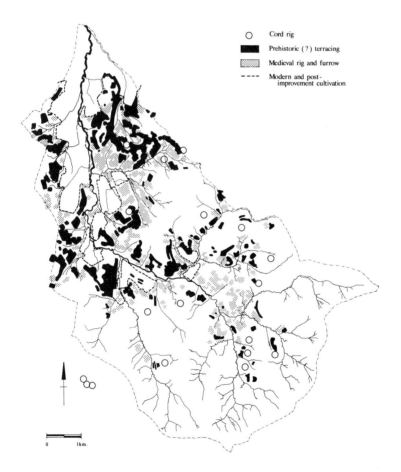

1.6: The distribution of cord rig Prehistoric terrain, Medieval rig and furrow and Post-Improvement agriculture in the Upper Bowmont Valley.

(ii) Early Bronze Age soil instability.

Tipping (1992) argued for an anthropogenic cause for this phase of soil erosion, but there are conflicts in the archaeological and pollen records that undermine this suggestion, and recent evidence to suggest that climatic change may have had a more substantial impact.

Burgess (1984) proposed that around 2000-1800 BC there was a pronounced shift in the focus of Early Bronze Age settlement in and around the Cheviot Hills, from lowlands such as the Milfield Basin, north west of the River Glen (Figure 1.1), where late Neolithic activity is relatively abundant (Harding 1981; Miket 1981), to the Cheviot uplands. The significance of this population movement is unclear from the archaeological record, however. Much rests on the veracity of Burgess' (1980, 1984) suggested Early Bronze Age date for construction of an unenclosed settlement at Houseledge, in the Harthope Valley above Wooler (Fig 1.1).

The difficulty of using the technique of house-platform construction as a chronological indicator is set out above. This is not to say, of course, that none of those recorded in the Bowmont Valley are of this early date, merely that in the absence of excavation there is no evidence to enable this view to be forwarded. What is clear is that many small platforms in the Bowmont Valley do relate to later periods where, by association or stratigraphy, they can be shown to be contemporary with, or later than, diagnostic house types of first millennium cal. BC date. The suggested early dating for the *floruit* of Hownam Law may, however, have implications in this regard.

The palynological record for the Bowmont Valley shows that, contrary to Burgess (1984; Burgess, Ovens & Uribe de Kellet 1981), the Cheviots were settled in the Neolithic period. Indeed, Mesolithic activity, from 4500 cal. BC, seems likely (Tipping *in press* b). Impacts of these occupations on plant communities were seemingly limited, and short-lived phases of apparent woodland regeneration occurred, suggesting, perhaps, sporadic settlement. Nevertheless, cereal cultivation is suggested after c2800 cal. BC at one pollen site in the valley, at Swindon Hill (Fig 1.1).

Crop-growing was sustained at Swindon Hill throughout the Bronze Age, ceasing at c900 cal. BC, but after c1900-1700 cal. BC there is recorded, at both Swindon Hill and Sourhope, a further increase in the extent of pasture, with renewed woodland clearance, possibly by grazing pressure. This pattern seems to be recorded at a number of southern Scottish pollen sites (Tipping *in press* c), and supports Burgess' (1984) proposal that this period marked the first significant population increase in the Cheviots. However, the apparent impact of these clearances on the natural woodland was still limited. No major clearance event is seen. Gates (1983, 110) saw the pattern of upland Bronze Age settlement as being one of farmsteads "scattered singly or in small groups separated by large expanses of apparently 'empty' landscape". This impression is supported by pollen analyses, with the 'empty' landscape occupied still by primary or secondary woodland surrounding 'oases' of cultivation and pasture.

Even in combination, the currently available records cannot determine settlement density in the uplands at this time. The geomorphological record might then suggest that human activities were more considerable than evidenced in either the archaeological or palynological data. This conclusion would provide real insight into weaknessses in our reconstructions of prehistoric settlement patterns, but there are problems in accepting this interpretation through, firstly, a strong competing hypothesis existing in climate change and, secondly, contradictions between proxy records.

New palaeoclimatic data from north west Europe combine to suggest a decisive and highly significant shift to wetter and colder climatic conditions after *c*2500-2400 cal. BC (Aaby 1976; Barber 1982; Charman 1990; Gear & Huntley 1991; Haslam 1987; Rothlisberger 1986; Tipping *in press* c; Walker 1966; Wigley & Kelly 1990). The climatic change seems to have been the final shift in a series of short-lived dry-wet oscillations in the preceding few hundreds of years. Large increases in precipitation would have a significant influence on fluvial discharges, frequency of flooding, channel movement and stream-bank erosion, and might be expected to destabilize soils on valley-sides, in the way the geomorphic record for the Cheviot Hills indicates.

This interpretation is made more likely by the agreement between the timing of the climatic deterioration and the onset of gravel deposition in the Cheviots. The Halter Burn radiocarbon date of *c*2450 cal. BC (above) is thought to relate most closely to the subsequent fluvial activity, and this date seems to accord better with the climatic record than with the evidence from palynology or archaeology for heightened anthropogenic activity. The earliest response to erosion of these rivers seems to pre-date the period of Early Bronze Age human impact.

Confident temporal correlation between independently-derived proxy records such as these is critical. Before a climatic cause is assumed for this phase, it must be assured that the timing of these events have been defined successfully. At present there is only limited confidence in the archaeological record; too few sites have been excavated, and even fewer dated. The pollen record is comprehensive for the Bowmont Valley, but we have necessarily to assume, for the moment, that these localized clearances are typical, in timing and scale, of other valleys in the Cheviots. Quantification of the absolute scale of clearance events is one of the outstanding problems confronting palynologists. The geomorphological record is based on two radiocarbon dates from separate valleys, and although each is a 'meaningful' date, synchroneity in this erosion phase has yet to be demonstrated; more data are urgently needed. In addition, it needs to be demonstrated, if a climatic cause is accepted, that other drainage basins responded to this increase in precipitation, and this is far from clear (Macklin & Lewin *in press*). The timing of the climatic deterioration itself is not firmly established; current estimates suggest a gradual transition over several hundred years, or abrupt but diachronous shifts at different sites.

There is, finally, no requirement for climate and human impact to be seen as mutually exclusive causes. It may have been, for example, that increasing precipitation led to extensions of the drainage network, through spring-head retreat and activation of now dry valleys and slopes, and that by doing so, areas of cleared land, that are in any case likely to erode more easily than afforested ground (Thornes 1987), became more prone to soil erosion.

Conclusions

The construction of a record of past soil erosion is difficult. The simplest method, the dating of river terrace deposits preserved on and above valley floors, is also one of the least direct measures of soil erosion. Nevertheless, linkages between these can be established from process studies, and in suitable drainage basins, demonstrated in palaeohydrological studies. Terrace deposits have the advantage of being able to generate a record of catchment-scale erosion, though frequently with a loss of information on precise sources of sediment.

In conjunction with other proxy records, this record can be a powerful tool for landscape reconstruction. The phase of Late Iron Age soil instablity identified in the Cheviot Hills is one example of the deeper insights into cause and response that are possible, and which encourage further and equally interesting questions to be asked. What were the governing factors behind the felling of woodland in such a concerted fashion? What socio-economic re-organization was needed to move beyond the small-scale assarting associated with individual farmsteads? How is this reflected in the archaeological record? Can the extent of woodland clearance in different drainage basins be 'calibrated' against the scale of soil disturbance? Was it deforestation that induced soil erosion, or the ploughing of land not previously broken? For how long were soil losses tolerated before terracing was constructed? Was terracing successful in inhibiting soil erosion?

The phase of Early Bronze Age soil instability raises different and more intractable questions concerning the temporal correlation of events, and the dependence on this procedure for the determination of possible cause.

Nevertheless, the work within the Bowmont Valley has illustrated that liaison between the social and natural sciences can lead to a truly 'interactive' landscape history.

It is essential to bear in mind that the sources of evidence that we have attempted to combine are not of like value. The palaeoenvironmental evidence is obtained locally, securely dated and derived from two independent data-sets – palynological and geomorphological. The archaeological evidence is obtained through the inexact 'science' (however rigorously pursued) of surface examination and only the final phases of activity on any site will be recognised. Without excavation, any site's full chronological span or spatial extent cannot be established, nor those materials likely to provide a *locally* dated archaeological sequence. *Faute de mieux*, dates for similar structures from a wide range of environments, some by no means identical to the Bowmont Valley, and some at a distance of as much as 150kms, have to be 'imported'. However, consistent observation of many hundreds of archaeological associations within the valley suggests that the *sequence* of events is broadly valid.

Correlation between the two sets of data are inevitably those of a 'best-fit'! Like all 'best-fit' exercises, it is makeshift and temporary and demanding of further testing. The products of this close collaboration are, however, intriguing, and serve as the basis for more comprehensive exploration. Both writers look forward to the stimulus that it is hoped this enquiry provides, leading to the accession of more, and better, data in understanding human impact on the Scottish environment.

Bibliography

Aaby B. 1976. Cyclic variations in climate over the past 5,500 yr reflected in raised bogs. *Nature* 163: 281-84.

Barber K E. 1982. Peat-bog stratigraphy as a proxy climatic record. In: Harding A F (ed). *Climatic Change in Later Prehistory*, 103-113. Edinburgh: University Press.

Barrow G W S. 1962. Rural Settlement in Central and Eastern Scotland - The Medieval Evidence. *Scottish Studies* VI, Part 2, 123-44

Bartley D D, Chambers C, Hart-Jones B. 1976. The vegetational history of parts of south and east Durham. *New Phytologist* 77: 437-68

Birks H J B, Birks H H. 1980. *Quaternary Palaeoecology*. Cambridge: University Press

Birks H H, Birks H J B, Kaland P E, Moe D (eds). 1988. *The Cultural Landscape - Past, Present and Future*. Cambridge: University Press

Brazier V, Ballantyne C K. 1988. Late Holocene debris cone evolution in Glen Feshie, western Cairngorm Mountains, Scotland. *Transactions of the Royal Society of Edinburgh: Earth Sciences* 80: 17-24

Brazier V, Whittington G, Ballantyne C K. 1988. Holocene debris cone evolution in Glen Etive, Western Grampian Highlands, Scotland. *Earth Surface Processes & Landforms* 13: 525-31

Brown A G. 1987a. Long-term sediment storage in the Severn and Wye catchments. In: Gregory K J, Lewin J, Thornes J B (eds) *Palaeohydrology in Practice*. Wiley & Sons, Chichester

Brown A G. 1987b. Holocene floodplain sedimentation and channel response of the lower River Severn, United Kingdom. *Zeitschrift für Geomorphologie* 31: 293-310.

Brown A G, Barber K E. 1985. Late Holocene palaeoecology and sedimentary history of a small lowland catchment in central England. *Quaternary Research* 24: 87-102.

Brown A G, Keough M. 1992. Holocene floodplain metamorphosis in the Midlands, United Kingdom. *Geomorphology* 4: 433-45

Burgess C. 1980. Excavations at Houseledge, Black Law, Northumberland, 1979, and their implications for earlier Bronze Age settlement in the Cheviots. *Northern Archaeology* 1: 5-12

Burgess C. 1984. The prehistoric settlement of Northumberland. In: Miket R & Burgess C (eds). *Between and Beyond The Walls*. John Donald, Edinburgh

Burgess C, Ovens M, Uribe de Kellet A. 1981. The ground and polished stone implements of north east England: a preliminary statement. *Northern Archaeology* 2: 6-12.

Burrin P J. 1985. Holocene alluviation in southeast England and some implications for palaeohydrological studies. *Earth Surface Processes & Landforms* 10: 257-71.

Chambers F M. 1993. *Climate Change and Human Impact on the Landscape*. Chapman & Hall, London

Charman D J. 1990. *Origins and development of the Flow Country blanket mire, northern Scotland, with particular reference to patterned fens*. Unpublished Ph.D. Thesis, University of Southampton.

Chorley R J. 1969. The drainage basin as the fundamental geomorphic unit. In: Chorley R J (ed). *Introduction to Fluvial Processes*. Methuen, London

Clapperton C M, Durno S E, Squires R H. 1971. Evidence for the Flandrian history of the Wooler Water, Northumberland, provided by pollen analysis. *Scottish Geographical Magazine* 57: 14-20.

Dickson J H. 1992. Scottish woodlands: their ancient past and precarious present. *Scottish Forestry* 47: 1-6.

Dodgshon R A. 1975. Towards an understanding and definition of runrig: the evidence for Roxburghshire and Berwickshire. *Trans.Inst.Brit. Geographers* LXIV:15-33

Dumayne L. 1992. *Late Holocene Palaeoecology and Human Impact on the Environment of North Britain*. Unpublished Ph.D. Thesis, University of Southampton.

Dumayne L. 1993. Invader or native? - vegetation clearance in northern Britain during Romano-British time. *Vegetation History and Archaeobotany* 2: 29-36

Fenton-Thomas C. 1992. Pollen analysis as an aid to the reconstruction of patterns of land-use and settlement in the Tyne-Tees region during the first millennia BC and AD. *Durham Archaeological Journal* 8: 51-62.

Gates T. 1983. Unenclosed settlements in Northumberland. In: Chapman J C, Mytum H C (eds). *Settlement in North Britain 1000BC-AD1000*. British Archaeological Reports, Oxford

Gear A J, Huntley B. 1991. Rapid changes in the range limits of Scots Pine 4000 years ago. *Science* 251: 544-47

Gregory K J. 1983. *Background to Palaeohydrology*. Wiley & Sons, Chichester

Gregory K J, Lewin J, Thornes J B. 1987. *Palaeohydrology in Practice*. Wiley & Sons, Chichester

Halliday S. 1982. Late prehistoric farming in SE Scotland. In: Harding D W (ed). *Late Prehistoric Settlement in South-East Scotland*. University of Edinburgh Department of Archaeology Occ.Paper No.8: 74-91

Harding A. 1981. Excavations in the prehistoric ritual complex near Milfield, Northumberland. *Proceedings of the Prehistoric Society* 46: 87-135

Harrison S. 1993. Solifluction sheets in the Bowmont Valley, Cheviot Hills. *Scottish Geographical Magazine* 109: 119-22

Harvey A M, Renwick W H. 1987. Holocene alluvial fan and terrace formation in the Bowland Fells, northwest England. *Earth Surface Processes and Landforms* 12: 249-57

Harvey A M, Oldfield F, Baron A F, Pearson G W. 1981. Dating of post-glacial landforms in the Central Howgills. *Earth Surface Processes and Landforms* 6: 401-12

Haslam C. 1987. *Late Holocene peat stratigraphy and climatic change - a macrofossil investigation from the raised mires of Europe*. Unpublished Ph.D. Thesis, University of Southampton.

Hill P. 1982a. Settlement and Chronology. In: Harding D W (ed). *Late Prehistoric Settlement in South East Scotland*. University of Edinburgh Department of Archaeology Occ.Paper No.8: 4-43

Hill P. 1982b. Broxmouth Hillfort Excavations, 1977-78 - An Interim Report. In: Harding D W (ed). *Late Prehistoric Settlement in South East Scotland*. University of Edinburgh Department of Archaeology Occ.Paper No.8: 141-88

Hughes J, Huntley B. 1988. Upland hay meadows in Britain - their vegetation, management and future. In: Birks H H et. al. (eds). *The Cultural Landscape - Past, Present and Future*. Cambridge: University Press

Hunter R F. 1962. Hill sheep and their pasture: a study of sheep-grazing in south-east Scotland. *Journal of Ecology* 30: 651-80

Innes J L. 1983. Stratigraphic evidence of episodic talus accumulations on the Isle of Skye, Scotland. *Earth Surface Processes and Landforms* 8: 399-403

Jacobson G J Jr, Bradshaw R H W. 1981. The selection of sites for palaeovegetational studies. *Quaternary Research* 16: 80-96

Jobey G. 1971. Excavations at Brough Law and Ingram Hill. *Archaeologia Aeliana* 4th Series 49: 71-94.

Jobey G. 1980. Greenknowe unenclosed platform settlement and Harehope cairn,

Peebleshire, *Proc.Soc.Antiq.Scot*.CX: 72-113

King J. 1960. Observations on the seedling establishment and growth of *Nardus stricta* in burned Callunetum. *Journal of Ecology* 48: 667-77

King J. 1962. The *Festuca-Agrostis* grassland complex in S.E. Scotland. *Journal of Ecology* 50: 321-55

King J, Nicholson J A. 1964. Grasslands of the forest and sub-alpine zones. In: Burnett J H (ed). *The Vegetation of Scotland*. Oliver & Boyd, Edinburgh

Lamb H H. 1981. Climate from 1000 BC to 1000 AD. In: Jones M, Dimbleby G W (eds). *The Environment of Man: the Iron Age to the Anglo-Saxon Period*. British Archaeological Reports, Oxford

Lewin J. 1981. *British Rivers*. George Allen & Unwin, London

Lowe J J, Walker M J C. 1984. *Reconstructing Quaternary Environments*. Pergamon, Oxford

Macinnes L. 1982. Pattern and purpose: the settlement evidence. In: Harding D W (ed), *Late prehistoric Settlement in South-east Scotland*. Dept. of Archaeology, Edinburgh University, Edinburgh

Macinnes L. 1984. Settlement and Economy: East Lothian and the Tyne-Forth Province. In: Miket R and Burgess C (eds). *Between and Beyond the Walls*. Edinburgh.

Macklin M G, Lewin J (in press) Holocene river alluviation in Britain. In: Douglas I, Hagedorn J (eds). *Geomorphology and Geoecology, Fluvial Geomorphology*. Zeitschrift fur Geomorphologie (Supplement)

Macklin M G, Needham S. 1992. Studies in British alluvial archaeology: potential and prospect. In Needham S, Macklin M G (eds). *Alluvial Archaeology in Britain*. Oxbow Press, Oxford

Macklin M G, Passmore D G, Rumsby B T. 1992. Climatic and cultural signals in Holocene alluvial sequences: the Tyne basin.In Needham S, Macklin M G (eds). *Alluvial Archaeology in Britain*. Oxbow Press, Oxford

Mercer R J. (ed) 1981. *Farming Practice in British Prehistory*. Edinburgh

Mercer R J. 1985. *Archaeological Field Survey in Northern Scotland* Vol.III: 1982-83. University of Edinburgh Department of Archaeology Occ.Paper No.11

Mercer R J. 1991. The highland zone: reaction and reality 5000 BC-2000 AD. *Proceedings of the British Academy* 76: 129-50

Mercer R J. 1991. The survey of a hilltop enclosure on Ben Griam Beg, Caithness and Sutherland District, Highland Region. In: Hanson W S, Slater E A. (eds). *Scottish Archaeology - New Perceptions*. Aberdeen University Press, Aberdeen

Miket R. 1981. Pit alignments in the Milfield Basin and the excavation of Ewart I. *Proceedings of the Prehistoric Society* 47: 137-46

Mourik, J M van. 1986. *Pollen profiles of slope deposits in the Galician area (N.W. Spain)*. Elsevier, Amsterdam

Muir J W. 1956. *The Soils of the Country round Jedburgh & Morebattle*. Memoirs of the Soil Survey of Great Britain. Her Majesty's Stationery Office, Edinburgh

Needham S, Macklin M G. 1992. *Alluvial Archaeology in Britain*. Oxbow Press,Oxford

Newson M D. 1981. Mountain streams. In: Lewin J (ed). *British Rivers*. George Allen & Unwin, London

Owen O A. 1992. Eildon Hill North, Roxburgh, Borders. In: Rideout J S,

Owen O A, Halin E. (eds). *Hillforts of Southern Scotland*. Historic Scotland Monograph No.1: 21-71

Passmore D G, Macklin M G, Stevenson A C, O'Brien C F, Davis B A S. 1992. A Holocene alluvial sequence in the lower Tyne Valley, northern Britain: a record of river response to environmental change. *The Holocene* 2: 138-47

Pearson G W, Stuiver M. 1986. High precision calibration of the radiocarbon time scale 500-2500 BC. *Radiocarbon* 28: 911-34

RCAHMS. 1956. *Roxburghshire: An Inventory of the Ancient and Historical Monuments.* Her Majesty's Stationery Office, Edinburgh

Piggott C M. 1949. The Iron Age Settlement at Hayhope Knowe, Roxburghshire; Excavation 1949. *Proc.Soc.Antiq.Scot* LXXXIII: 45-67

Ralston I B M. 1983. High altitude settlement on Ben Griam Beg, Sutherland. *Proc.Soc.Antiq.Scot* 113: 636-38

Richards K S, Peters N S, Robertson-Rintoul M S E, Switsur V R. 1986. Recent valley sediments in the North York Moors: evidence and interpretation. In: Gardiner V.(ed). *International Geomorphology Part I.* Wiley & Sons, Chichester

Rideout J R, Owen O A. 1992. *Hillforts of southern Scotland.* AOC (Scotland) Ltd., Edinburgh

Robinson M A, Lambrick G H. 1984. Holocene alluviation and hydrology in the Upper Thames basin. *Nature* 308: 809-14

Rothlisberger F. 1986. *10,000 Jahre Gletschergergeschichte der Erde.* Sauerlander, Aarau

Sanderson D C W, Placido F, Tate J O. 1988. Scottish Vitrified Forts: The results from six study sites. *Nuclear Tracks* 14: 307-16

Shotton F W. 1978. Archaeological inferences from the study of alluvium in the lower Severn-Avon valleys. In: Limbrey S, Evans J G (eds). *Man's effect on the landscape: the Lowland Zone.* Council for British Archaeology, London

Smith C. 1988-89. Excavations at Dod Law West hillfort, Northumberland. *Northern Archaeology* 9: 1-55

Starkel L, Gregory K J, Thornes J B. 1991. *Temperate Palaeohydrology–Fluvial Processes in the Temperate Zone during the last 15 000 Years.* Wiley & Sons, Chichester

Stevenson R B K. 1941. Medieval dwelling sites and a Primitive Village in the Parish of Manor, Peeblesshire. *Proc.Soc.Antiq.Scot* LXXV: 92-115

Stuiver M, Pearson G W. 1986. High precision calibration of the radiocarbon time scale 1950-500 BC. *Radiocarbon* 28: 805-38

Thornes J B. 1987. Models for palaeohydrology in practice. In: Gregory K J, Lewin J, Thornes J B (eds). *Palaeohydrology in Practice.* Wiley & Sons, Chichester:

Tipping R M. 1992. The determination of cause in the generation of major prehistoric valley fills in the Cheviot Hills, Anglo-Scottish Border. In: Needham S, Macklin M G (eds). *Alluvial Archaeology in Britain.* Oxbow Press, Oxford

Tipping R M (in press *a*). Fluvial chronology and valley floor evolution of the upper Bowmont Valley, Borders Region, Scotland. *Earth Surface Processes and Landforms.*

Tipping R M (in press *b*). *The Bowmont Valley. The Holocene Evolution of an Upland Scottish Landscape.* Dept. of Archaeology, University of Edinburgh.

Tipping R M (in press *c*). Holocene evolution of a lowland Scottish landscape: Kirkpatrick Fleming. I. Peat- and pollen-stratigraphic evidence for raised moss development and climatic change. *The Holocene.*

Tipping R M (in press *d*). Holocene evolution of a lowland Scottish landscape: Kirkpatrick Fleming. II. Regional vegetation and land-use change. *The Holocene.*

Tipping R M, Halliday S P. 1994. The age of alluvial fan deposition at a site in the Southern Uplands of Scotland. *Earth Surface Processes & Landforms* 19: 333-48.

Topping P. 1989. Early cultivation in Northumberland and the Borders. *Proc.Prehist.Scot* LV: 161-80

Triscott J. 1982. Excavations at Dryburn Bridge, East Lothian 1978-79. In: Harding D W (ed). *Later Prehistoric Settlement in SE Scotland.* University of Edinburgh Department of Archaeology Occ.paper No.8: 117-24

Turner J. 1979. The environment of northeast England during Roman times as shown by pollen analysis. *Journal of Archaeological Science* 6: 285-90

Veen M van der. 1992. *Crop Husbandry Regimes - An Archaeological Study of Farming in Northern England 1000BC-AD500*. J R Collis Publications, Dept. of Archaeology and Prehistory, University of Sheffield

Walker D. 1966. The late Quaternary history of the Cumberland Lowland. *Philosophical Transactions of the Royal Society of London* B251: 1-210

Welfare A. 1992. *The Breamish gravel quarry, Powburn, Northumberland : a southern extension - an assessment of the potential archaeological constraints*. Heritage Site and Landscape Surveys Ltd.

Wigley T M L, Kelly P M. 1990. Holocene climatic change, ^{14}C wiggles and variations in solar irradiance. *Philosophical Transactions of the Royal Society of London* A330: 547-60

Wilson D H. 1981. *Pollen Analysis and Settlement Archaeology of the First Millennium BC from North -east England*. Unpublished M.Sc. Thesis, University of Oxford.

Young A. 1972. *Slopes*. Oliver & Boyd, Edinburgh

Acknowledgements

The authors are pleased to acknowledge the support of Historic Scotland in the funding of this work, and thank in particular Noel Fojut, Patrick Ashmore and Gordon Barclay for their unstinting support. Roger Mercer wishes to thank Jack Stevenson and Stratford Halliday for reading his part of this text and for their valued comments. Carole Buglass performed miracles in inter-digitating text presented in different styles on different sized diskettes! Richard Tipping is grateful to the University of Edinburgh Development Trust and the British Geomorphological Research Group for continuing financial support. The NERC Radiocarbon Dating Committee supported the funding of nearly all the radiocarbon dates discussed above; Historic Scotland funded those from Yetholm Loch at the SURRC laboratory. We are grateful for the care and commitment of both the NERC and SURRC dating laboratories.

We also thank Christopher Smout for the opportunity to speak at the second Battleby Conference, and for the invitation to prepare this text.

Chapter 2

Field-Systems, Rig and Other Cultivation Remains in Scotland: The Field Evidence

Piers Dixon

The aim of this chapter is to review previous archaeological survey work on Scottish medieval or later field-systems, to define the main types of features attributable to them, and to assess the evidence of several case studies from recent work by the Royal Commission on the Ancient and Historical Monuments of Scotland (RCAHMS) in the light of current understanding.

The Archaeological Background

The physical remains of medieval and later field-systems have been neglected by the archaeological community in much the same way as the settlements of these periods. As with settlements, it was an historical geographer, Horace Fairhurst, who made the first steps in their analysis with his study of the plough-rig within the ring-dyke that surrounded the township of Rosal in Sutherland (Fairhurst 1968).

Previous interest in these field-systems was indirect and centred mainly on cultivation-terraces and only on rig in-so-far-as they related to the terracing (Graham 1939; Stevenson 1947). The RCAHMS's interest in medieval and later field-systems stemmed from the work on cultivation-terraces in the Border counties of Scotland and benefited from the vertical aerial photographs that were produced by the National Aerial Survey immediately after the Second World War. It was in this context that the field-system at Old Thornylee was published in the Peeblesshire Inventory (RCAHMS 1967), showing how an earlier field-system had been moulded by medieval and later rig cultivation (Pl. 2.1). However, the potential that this survey revealed for understanding the field-systems of the medieval and later periods was limited by the available ground-survey techniques, which made the mapping of widespread field remains impractical.

Parry, in his research on marginality in the Lammermuirs, attempted to create a typology of plough-rig in south-east Scotland (Parry 1976), building on the seminal work of Bowen who identifed 'Broad' and 'Narrow' plough rig from his work in England (Bowen 1961). Parry tested this paradigm by measuring the wavelength and amplitude of the rig in selected locations and concluded, not surprisingly, that there were two types of rig: Type 1, the broad high-backed sinuous rig over 4.5m in breadth, and Type 2, a generally narrower rig, less

Pl. 2.2: House-plots, or tofts and crofts at Midlem, Roxburghshire
(copyright RCAHMS RX/5197).

Pl. 2.1: Old Thornylee, Peeblesshire showing broad rig overlying and moulding cultivation-terraces
(copyright RCAHMS PB/2334).

than 4.5m and of low profile. A third type (Type 3), he defined, was essentially Type 2 over Type 1, that is modifying an existing broad rig system. He also used documentary sources to place Type 2 firmly in the Improvement period of the later eighteenth to early nineteenth centuries, prior to the introduction of underground drainage in the nineteenth century. The origins of Type 1 were asserted to be in the High Medieval period, produced with the heavy mould-board plough that was introduced by the monastic estates in the early twelfth century, although no clear-cut evidence for this introduction has been presented. On the other hand, Parry associated Type 2 with the introduction of Small's swing-plough in 1767 into Berwickshire, but the move to straighten and level ridges has its origins in the early eighteenth century and cannot be exclusively attributable to the swing-plough. It is illustrative of the lack of work by archaeologists that this has remained the situation until recently.

In arable land a different technique is needed to record the ridges as they have long since been levelled. Oblique aerial photographs were used by Pollock in his study of the landscape of the Lunan Valley in Angus where he was able to map the cropmarks created by the furrows of rig in arable ground. Pollock was able to confirm that the rig revealed in this way conformed with that which survived in a woodland plantation and was of a broad sinuous form that Bowen would have recognised as 'Broad' rig (Parry Type 1).

However, archaeological excavation is required if we are to date the formation of the ridges, whether in the uplands or in the lowlands. Indeed, Pollock's excavations in the Lunan Valley indicated that the broad rig post-dated, in one instance at least, a group of medieval corn-drying kilns, giving a rough *terminus post quem* for their formation (Pollock 1985), whilst a recent excavation at North Straiton in Fife produced twelfth- and thirteenth-century pottery from the fill of a shallow ditch of a field-system that was superseded by rig also containing medieval pottery (DES 1987). In these two instances the date of formation of the rig appears to be late medieval, but what of the ditched field-system of North Straiton of which we know little and understand even less? What is needed is the excavation of sample rig from a variety of locations, selected, preferably, from completely mapped field-systems so that the spatial context of the rig is known.

New technology, with the advent of the Electronic Distance Measurer in the 1980s, has transformed the possibilities of mapping medieval and later field-systems on a substantive scale, as has been demonstrated by the RCAHMS in North-east Perth (RCAHMS 1990), and on Waternish, Skye (RCAHMS 1993a). However, a map of the archaeological remains, unlike an estate map, which is essentially a snap-shot of the relevant features at the date of survey, is a record of the physical relics of events covering hundreds, if not thousands, of years. Therefore one should not expect a total coincidence of one with the other, but the estate map should help elucidate some of the features on the archaeological survey and provide a *terminus* for the points of co-incidence and a starting point for further research. Furthermore for many parts of Scotland, the archaeological map will be the only indication of past activity, since documentation is patchy, particularly for events that pre-date the later eighteenth century when documentation becomes plentiful.

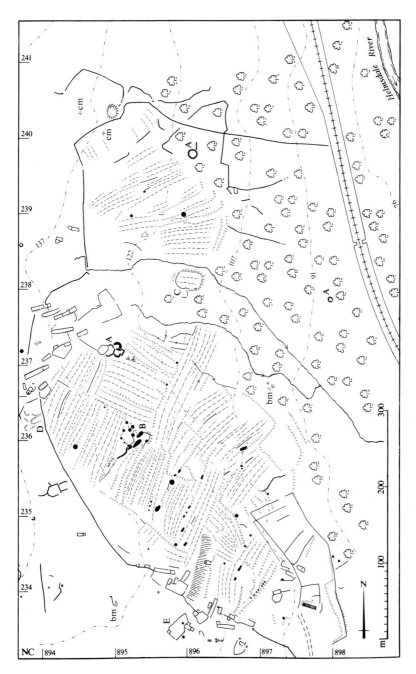

2.1: Plan of a head-dyke enclosing a broad-rig system at Learable, Sutherland.
Note the terraces (copyright RCAHMS)

The Documentary Background

Our notions of the organisation of Scottish medieval and later field-systems, and in particular the infield-outfield system and 'runrig', are derived from the work of historical geographers, based largely upon post-medieval sources and estate maps in particular. The well-known model for infield-outfield, that is outlined by Whittington (1973), is based upon a survey of an Aberdeenshire farm of 1760 (cf. Wilson 1902) and relates well to the descriptions to be found in the eighteenth century treatises on agriculture and in the *Statistical Account* (Stat. Acct., 1791-9). In essence, infield-outfield is a rotational system of cropping land, based upon an infield core, regularly manured and cropped and a wider area of ground that was temporarily taken in and cropped as required. At the same time the work of Dodgshon has contributed substantially to the definition of the runrig organisation of Scottish pre-Improvement field-systems (e.g. Dodgshon 1980). Runrig was the sub-division of arable land amongst the various landholders of a fermtoun or township, such that an individual tenant held land, usually based upon the plough-rig, intermixed with that of his co-tenants throughout the lands of the toun. This required a communal system of management.

The origins of runrig and of infield-outfield are less well understood. Barrow, working exclusively on medieval sources, could find no references to infield or outfield in the twelfth and thirteenth centuries (Barrow 1962); indeed the terms only come into use from the fifteenth century (Dodgshon 1980). Although Barrow found a distinction in the way in which the cultivated lands of a township were described in the south-east (in numbers of bovates and carucates) from that with which they were described north of the Forth (davochs or named carucates) in the twelfth and thirteenth centuries, in both areas the basic units of landholding were the rig (Latin: *reia*) or the acre, and landholdings were frequently sub-divided into rigs or acres, and spread in a number of locations throughout the farm or township. Indeed, one isolated example from Fife at the turn of the thirteenth century refers to a landholder having every fifth rig of the toun of Ballebotlia. Whilst this may be an early example of sub-divided fields based upon the rig, the term runrig, as with infield-outfield, does not appear before the fifteenth century (Dodgshon 1980). It is thus not entirely clear whether the system developed before the end of the medieval period, or not.

Be that as it may, the basic unit, the rig, was already documented as a common feature in townships in the medieval period. Furthermore, the furlong (L: *cultura*), which appears to be an agglomeration of rigs, sometimes called shotts or butts and other names, also makes its appearance, but essentially as a topographic unit rather than a cropping unit. The overall fields (L: *campus*) of a township may be composed of one or more of these. Whilst these appear to be the basic units of medieval and later field-systems, it is not evident, at present, when they were first introduced or developed, but from the evidence presented by Barrow they appear to be already well-established in the later twelfth and thirteenth centuries in the southern and eastern counties of Scotland.

Dodgshon (1980) has argued cogently that infield-outfield and the fiscal assessment of cultivated land in a township (e.g. the number of bovates and caru

Pl. 2.4: A broad rig-system at Balquhatson, Stirlingshire. Note the rectilinear hedged enclosures overlying the rig, which are themselves abandoned (copyright RAF 541/A/441 No.4267).

Pl. 2.3: Broad rig enclosed by earthen banks which overlie the rig at Murrayfield Plantation, Fife. Note also the grooving along the centre of the rig (copyright RCAHMS KR/460).

cates, or latterly husbandlands) are inter-related. As new land was taken in under the pressure of population from the end of the medieval period, outfield became more regularly cropped, but as the new land was technically outfield, it remained in excess of the medieval assessment of land. However, although runrig is not documented until this period of expansion, it seems unlikely that it was created especially to deal with the new lands.

However, such documentary and theoretical discussions often seem to me, as someone who works with the physical remains of these activities, to be somewhat detached from my own frame of reference. Whilst it is possible to recognise rig, is it possible to recognise infield or outfield from its appearance on the ground? This seems unlikely, since both will be rigged. There certainly seems no hope of identifying runrig, since it is organisational, and rig in itself does not mean that there is a runrig system, since this depends upon there being more than one tenant.

On the other hand, basic questions of definition are needed. What is a field-system and what is the physical remains of it? Is there regional variation? How do these systems compare with the documentary evidence and the infield-outfield paradigm?

This chapter concentrates upon the cultivated lands, as opposed to the settlement remains, and leaves transhumance and the management of cattle and sheep as an unspoken, but essential, element in the business.

Definitions

The pun in the title begs the question as to what we mean when we use the term field-system. The modern notion is that a field is enclosed, that is a fenced area within which cultivation, usually with some form of crop rotation, takes place. A field-system is merely several of these units, forming the cultivated lands of a farm, based upon a farmsteading. This type of field-system became the norm from the Agricultural Revolution of the late eighteenth and early nineteenth centuries, otherwise known as the Improvements, since it entailed the complete abandonment of the communal runrig system. Medieval and later cultivation remains, and in particular rig-systems, were not enclosed and the term rig-system may be more descriptively appropriate. On the other hand some rig-systems were enclosed within a ring-dyke or head-dyke, in which case the term field-system may be appropriate, but unlike a modern farm the cropping-units were not physically sub-divided or fenced. Thus field-system is simply used in this paper to describe any area of ground which displays evidence of cultivation-remains. Ideally this should be related to a settlement, the hub of the system, but the exigensies of survival mean that this is not always possible.

For the purposes of this chapter, I will outline the different types of structure associated with medieval and later field-systems. For convenience these are divided into two, as follows:

Enclosure
An enclosure is a fenced piece of ground, which allows exclusive activity to take place within it. Its size, location and the type of fence that is employed will vary

Pl. 2.5: A park pale at Buzzart Dykes, Perthshire
(copyright RCAHMS A 55554/CN).

with its function. Thus enclosed areas include farmyards or house-plots, garden plots, animal pens, crofts, head-dykes and ring-dykes, fields and boundary dykes.

a. Small enclosures – tofts and crofts, yards and gardens, pens and stock enclosures

The farmstead usually includes a yard, which, indeed, is viewed archaeologically as a defining feature (e.g. Fig. 2.6). The farmstead or toft of medieval documents is often associated with a croft in the possessions of a tenant in the south-east (e.g. on the Coldingham Priory estates, cf. Raine 1852). The croft was a piece of cultivated ground of an acre or two in size for the exclusive use of the tenant, separate from the runrig arable, typified today by the crofting-township of the Highlands, but in earlier periods by the village plots of the south-east as at Midlem (Pl. 2.2).

A variety of animal pens or plots for gardens occur, the latter usually in relation to settlement. Sheep-pens and stock enclosures are, in general, a post-medieval feature, so far as it is possible to tell in the present state of knowledge, associated with the development of sheep-farming (see Fig. 2.2) or cattle-droving from the seventeenth century. These may encompass large areas of pasture and may, as on Waternish, Skye, be referred to as parks (RCAHMS 1993a). The enclosing walls may be of turf, turf and stone, stone, or earthen, and stand to as much as 1.2m in height, a sufficient barrier to stock. Where there are garden plots, they generally reveal themselves by the deeper soil of the interior, due to the spreading of midden material (e.g. Waternish, Skye).

b. Fields

Earthen or stone dykes may be found enclosing blocks of rig. These may vary in plan from the sub-circular or irregularly-shaped to the rectilinear. The former are to be found in a number of localities from Galloway and Skye to Perthshire and Fife. Some are quite evidently attempts to enclose, piecemeal, areas of rig, presumably as a first step towards improvement, often with the banks overlying the rig around the fringes of the fields (Pl. 2.3). How early these should be dated remains to be tested, but, in some cultivation, continued and in others ceased according to circumstance. However the irregularity of the fields, enclosing former or current areas of rig, reveals their limitations to improved agriculture, and they were subsequently abandoned in course of the Improvements. The latter, or rectilinear type of field, is usually of later eighteenth to nineteenth century date, and is associated with the progress of Improvement, but has gone out of use since then (Pl. 2.4 & Fig. 2.4).

c. Head-dykes and ring-dykes, assarts, park pales and boundary dykes

Many medieval or later field-systems are characterised by a head-dyke or ring-dyke that encloses the main area of arable of a farm within an earthen bank or bank and ditch. It is difficult to say how early this form of enclosure was employed. Many of the recorded head-dykes are demonstrably late features (e.g. Learable, Sutherland, Fig. 2.1), indicative perhaps of pressures on the common grazings in the post medieval period, on the other hand this relationship is what should be expected if townships were expanding in the post medieval period and new head-dykes had to be constructed as on Skye (RCAHMS 1993a). Some, however, are much earlier and medieval in date. Recent work at Southdean in the Forest of Jedburgh has identified several forest steads and found that their lands were enclosed within banks and external ditches, designed to limit access to intruding deer, but not egress (RCAHMS 1994). These are called assarts in the terminology of medieval hunting reserves, that is to say pieces of ground specially licenced for cultivation and settlement within the hunting forest, whose bounds could be defined by a ditch (Guilbert 1983).

Another form of enclosure also found in the hunting reserve is the park pale (Guilbert 1979, RCAHMS 1990), except here no settlement is intended and the pale comprises a bank and ditch, this time with the ditch on the inside and a palisade was set into the bank, in order to trap deer within its bounds (Pl. 2.5).

Boundary dikes or ditches are another occasional feature of the settlement landscape. Farm and estate boundaries were usually defined by natural features such as burns, or prominent features, but as with assarts, a boundary may need to be defined, particularly in the case of a dispute. What may be a medieval assart ditch was found at Southdean, where it was superseded by some of the medieval fields. Combined with natural features such as burns, it defines an area of several square kilometres (RCAHMS 1994).

Rig

The main feature that dominates any discussion of medieval and later field-systems are various types of ridge and furrow or rig. However, it is also recognised that ridging is an old form of preparation of the soil for cultivation which has its origin in the prehistoric period. Ridging both improves drainage in the ridge, by providing ditches to take run-off water, and the temperature of the soil on the ridge, thus aiding germination of the seed-corn (Lerche 1986). It is a method of soil preparation found widely in northern Europe, but also in pre-Conquest America. It can be created both with a spade or a plough, but it is central to the substance of this paper that there is a difference between the ridge created in these two ways, but of course it is possible for a plough ridge to be reworked with a spade or *vice versa*, and for a plough (or ristle on the north-west coast) to be used for the initial breaking of the sod, preparatory to building a ridge with a spade.

Before outlining the main types of medieval and later rig, it should be appreciated what prehistoric ridging was like. The term 'cord-rig' has been coined to describe a prehistoric form of rig cultivation. The rig is very narrow, generally about 1.25m in width, and relatively short in length, usually less than 100m. It is found in groups of contiguous ridges and blocking of these groups sometimes occurs. The formation of cord-rig remains unresolved and such are the slightness of the ridges, possibly due to soil degradation, that it will not be easy to do so. However, it is important to distinguish this type of rig from later episodes.

The types of rig attributable to the medieval and later periods include the following, building upon the work of Bowen (1961) and Parry (1976):

Pl. 2.7: Broad rig oblique to the slope, creating assymetrical rig at Sheriffmuir, Perthshire (copyright RCAHMS B10253).

Pl. 2.6: A broad rig field-system surviving as a crop-mark at Leuchars, Fife. (copyright RCAHMS F/6567).

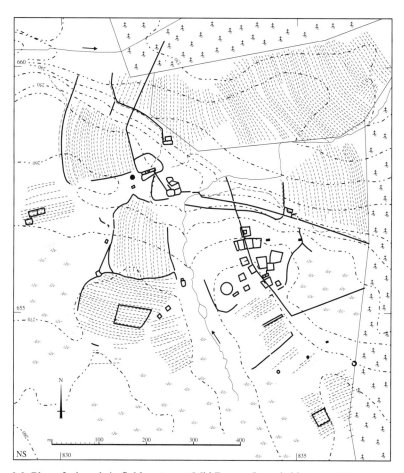

2.2: Plan of a broad rig field-system at Mid Bracco, Lanarkshire (copyright RCAHMS)

a. Broad or reverse-S rig (Parry Type 1):
This is rig that is generally in excess of 4.5m and up to perhaps 20m in width
(the rigs displayed by the cropmarks of rig-systems at Leuchars in Fife are up to
20m across, Pl. 2.6), typically displaying a curve in the shape of a reverse-S and
extending to about 250m in length. The formation of broad ridges has been discussed in detail by Hall, based upon his work in the East Midlands (Hall 1982).
Suffice it to say, that broad rig is intentionally formed, depending upon the use
of a fixed mould-board plough and the continual turning of the sod inwards as
the plough-team works around a strip of land.

Rigs are usually contiguous and grouped into blocks, the furlongs, shotts or butts of the documents. Where furlongs lie close together, they may be end-on to one another with a headland between, or at right angles blocking its neighbour, in which case the headland is itself cultivated as part of the neighbouring furlong. Because of the needs of drainage, the rig runs across the contour or downslope, but the exigencies of slope sometime dictate that part of a furlong may run oblique to the slope. This may lead to soil slippage and the creation of asymetrical rig (Rosal, Sutherland or Sheriffmuir, Perthshire, Pl. 2.7). Or as at Old Thornylee, where there were earlier cultivation terraces, the rig may run along the contour, but since the terraces are rarely designed with the medieval plough in mind, this can lead to the stepping or moulding of the terraces (e.g. Braemoor Knowe, Roxburghshire, Pl. 2.8).

Well-used and well-formed examples have a distinctive 'high-back' with the ridge rising gently from the furrow to the crest of the ridge and the characteristic reverse-S shape (e.g. Sheriffmuir, Perthshire, Pl. 2.9), which was the result of the need to keep the plough biting the ground to the end of the ridge as the large plough-team was turned on to the headland. Not all such rig is particularly well defined, presumably because of the short-lived nature of the episode of ploughing. In these cases the furrow-groove is the most visible part, since the ridge will be poorly formed (Pl. 2.10 and RCAHMS 1994). Extensive grooved rig may be seen in the Cheviots, but it is also found elsewhere.

Broad rig is distributed from Sutherland in the north to the Border counties in the south with a pronounced eastern bias, although it is found in lowland Ayrshire and in Lanarkshire. It is a pre-Improvement type of cultivation that finally went out of use in the Improvement period – i.e. by about 1800 – but whose origins lie somewhere in the medieval period following the introduction of the heavy mould-board plough, perhaps with the advent of the reformed monasteries or the Anglo-French in the twelfth and thirteenth centuries, but this remains unproven. The implement that was used to create it was the Old Scotch Plough, which is referred to in the Reports of the Board of Agriculture at the turn of the nineteenth century and is described in eighteenth century agricultural treatises (Jirlow and Whitaker 1957). It was drawn by teams of oxen or horses or a mixture of both.

Broad rig is the classic fossil of medieval plough cultivation, common also in England and other north European countries such as Denmark (Lerche 1986), where it also can be up to about 20m in breadth. As in England, the precise date of its introduction is not well defined and excavation of a range of examples is needed. It is possible that its introduction was gradual, spread over centuries, perhaps with the south-east being the first area to adopt it, but evidently some areas never did.

A sub-group of broad rig is sub-divided rig. This is broad rig that has been sub-divided in the interests of the improvement of the soil and drainage as advocated as early as the first quarter of the eighteenth century (Parry 1976). Such rig may be recognised by the alternating shallower and deeper furrows.

b. Narrow curving rig (Pl. 2.11).
This is a form of plough rig that is to be found in Galloway and possibly in

Pl. 2.8: A broad rig-system at Braemoor Knowe, Roxburghshire. Note the asymetrical rig and the terracing (copyright RAF 106G/SCOT/UK 121 No.3264).

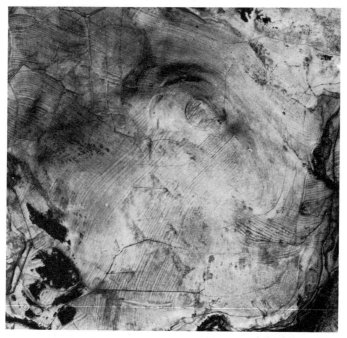

Pl. 2.10: A broad rig-system at White Hill, Roxburghshire. Note the grooved rig furlongs on the east and to the north-west of the hill (copyright RCAHMS B78663).

Pl. 2.9: A broad rig furlong at Sheriffmuir, Perthshire. Note the reverse-S shape and the clearance heaps in the furrows and the grooving along the crest of some of the ridges (copyright RCAHMS A 55463).

other west coast areas such as Argyll, but may prove to have wider distribution when sufficient survey work has been done. The rigs are contiguous and quite narrow at between about 2m and 4.5m. Where the terrain is gentle and unbroken, the rig forms furlongs and displays a reverse-S shape, as does broad rig, but in broken terrain it can be curvilinear with a distinctive pinching at the ends where a headland may be visible. It may be better viewed as a sub-type of broad rig formed in response to a more broken topography. The dating of this type of rig can only be identified as a pre-improvement type, but not at present its antiquity.

c. Narrow straight rig (Parry Type 2)
This is an improved type of rig, dating to the later eighteenth and nineteenth centuries, generally less than 4.5m in width, but rarely much less, straight and low backed (Pl. 2.12). It is distributed widely in Scotland, wherever improved agriculture was practised, prior to the introduction of underground drains in the later nineteenth century. It is to be distinguished from sub-divided broad rig in that it was generally formed anew. It was formed, it is argued (Parry 1976) by the lighter swing plough that became popular in the late eighteenth century.

d. Lazy-beds (feannagan)
This is predominantly the product of spade-digging (the cas-dhireach on the north-west coast) or the use of the cas-chrom (Fenton 1963). Lazy-bed ridges are raised by digging out the soil from the furrows on either side to build a deeper seed-bed for growing crops, the ground having been broken by the use of the ristle or perhaps an ard. The beds range from about 2m to 5m wide, but the furrows may be even wider than the ridge where the soil is shallow (Pl. 2.13). The edges of the ridges are sharp in the first instance, although they weather to a steep-sided ridge and stand up to some 0.45m in height. They are often sinuous or wiggly, and may be split or turn sharp angles, but they may be contiguous as well, depending upon the terrain. Unlike plough-rig they are found on slopes far too steep for a plough-team to negotiate.

The distribution of lazy-beds is mainly confined to the north-west, but they are also found in Argyll and Galloway, and small potato patches are found in upland areas of the south-east, such as the Cheviots, in the vicinity of farms. The origin of lazy-beds is obscure, but it is evidently a pre-Improvement type, which persisted in some crofting areas of the north-west until the twentieth century.

e. Stone clearance and strip-fields
Incidental to all these types of ridging is the clearance of stone where it is needed. This also has a long history as far back as the neolithic period. With rig-systems most clearance lies in the furrows or ditches between the rig so that dumps of stone will either take the form of debris in the furrow or of a low stony bank (Pl. 2.9), but individual dumps may also occur (Edwards 1980). These are generally distinguished from prehistoric clearance episodes by their location and linear arrangement, prehistoric clearance being formed of random circular mounds and are usually well covered with vegetation.

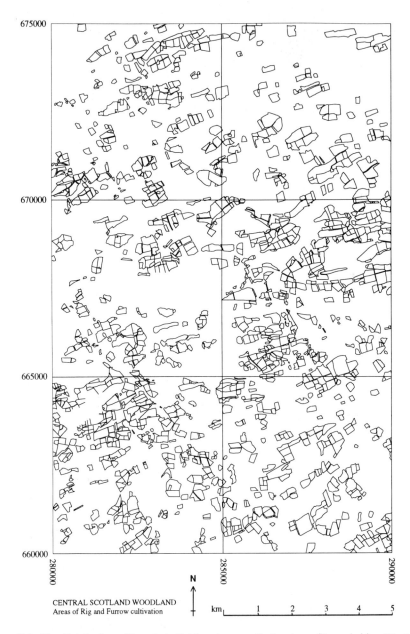

2.3: The distribution of broad-rig field-systems on the borders of Lanarkshire, West Lothian and Stirlingshire derived from National Aerial Survey photographs for the Central Scotland Woodlands Afforestable Land Survey (copyright RCAHMS)

Pl. 2.12: Narrow straight rig at Broadlee Hill, Selkirkshire. Note the truncated broad rig and the traces of grooved rig on the left side of the photograph (copyright RCAHMS B 17160).

Pl. 2.11: Narrow curving rig at High Eldrig, Wigtownshire. Note the lazy-beds adjacent to the farmstead (copyright RCAHMS WG/1254).

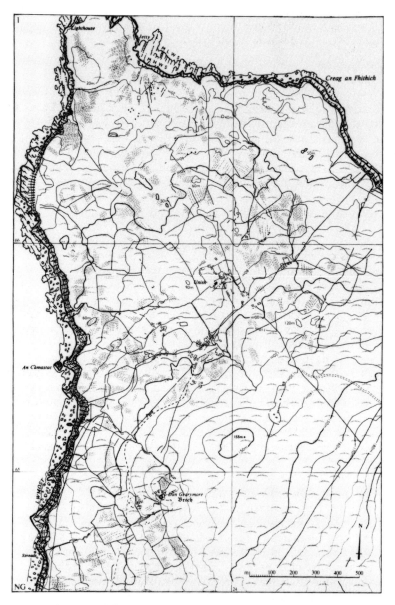

2.4: Plan of a lazy-bed field-system at Unish, Skye with sub-circular enclosures and
19th century rectilinear fields with smoothed area due to modern ploughing
(copyright RCAHMS)

Linear clearance in association with ridging may lead to strips being out-lined, as at Ranageig in Perthshire, where a series of roughly parallel banks of dumped stone defined the strips. Here there is little ridge development, but the shape of the strips would suggest that that this was the intention. The dating of these strips is unresolved at present, although there are buildings associated with the Ranageig site which may help provide it (RCAHMS 1990).

Case Studies

1. A broad rig field-system at Mid Bracco, Lanarkshire (Fig. 2.2)

Mid Bracco occupies a piece of hill-pasture on the eastern limits of Lanarkshire. The rigged ground occupies about 20ha in three main blocks, some having been lost in the plantation to the east, and is typical of the extent of rigged land to be found in association with pre-improvement farmsteads on the borders of Strathclyde, Lothian and Central Regions around Slamannan, Blackridge and Shotts (Fig. 2.3). The gaps between the rigged land here are generally occupied by boggy ground, unsuitable for cultivation and a major limit on settlement even today.

The rig at Mid Bracco displays many of the features typical of broad rig, in width, shape and length and is well-formed with a high back. The rig here as elsewhere is complex, revealing changes and alterations in the lay-out of the furlongs of rig. For example, to the north-east of the farmstead, a sharp bend in the rig suggests either that two furlongs end-on to one-another were combined or that the original furlong has been extended. Blocking, the lay-out of a furlong at right-angles to another, is also evident here as a curious small furlong sur-rounded on three sides by its neighbour. In places the rig has been superseded by earthen banks, which have themselves gone out of use. There is more than one phase to this process. Some of the banks cut across the rig, suggesting that this was enclosure for sheep or cattle, but elsewhere there has been piecemeal enclosure of the rig, with cultivation continuing, as in the field to south of the farmstead. Significantly there is no overall head-dyke and for the most part the rig just runs to the edge of boggy areas or the rough hill-pasture. To the south-east of the farmstead there is a group of rectilinear sheep-pens, possibly of the same period as the enclosures which were built across the rigged ground.

2. A lazy-bed field-system at Unish, Waternish, Skye (Fig. 2.4)

This lazy-bed system at Unish includes some lazy-beds which are contiguous, while others are not and they are to be found both in unenclosed patches and within sub-circular enclosures. The central area of the township has been culti-vated in recent times and presents a smooth greensward with the fragments of rectilinear fields, defined by earthen banks, presumably of nineteenth century date and indicative of an episode of improved farming. Several phases of head-dyke are apparent; this may reflect the exhortation of the Macleod's to their tenants through their tacks in the later eighteenth century to enclose their farms, as well as the expansion of population during the later eighteenth century (RCAHMS 1993a). The sub-circular fields, in the present state of knowledge, are a peculiarity of Waternish. They appear to be peripheral to the main expanse of lazy-beds and may indicate the piecemeal enclosure of small plots of arable, again under pressure of new tacks from the estate. This is confirmed to some extent by the estate map of 1790 (Pl. 2.14), wherein many of these enclosures are separately itemised fields, presumably arable, and must have been separate from the rest of the common arable.

The antiquity of this lazy-bed system remains to be tested. This township may be traced to the early eighteenth century with some confidence from the documentation and from the surviving structures, including a tacksman's house of that date. Whether the township was farmed in runrig is unclear. Other townships of Waternish had several tenants in the late seventeenth century, and lazy-bedding does not preclude it, even if it is difficult to understand how arable shares were allocated with rig that is not in any way contiguous or equal in area. If there was outfield it is not possible to identify it, unless the sub-circular enclosures served in this capacity.

3. A narrow-curving rig-system at Stroan, Kirkcudbrightshire (Fig.2.5)

At Stroan in Galloway the rig is found in a number of discrete areas on hillocks confined by a sea of bog. Occasionally the rig may be seen to predate the field-banks of the sub-circular fields but, elsewhere, cultivation continued within them and rig has built-up against the dykes. These fields were still being created at the north-east side of the site where clearance is partial when the farm was abandoned. With such a dispersed lay-out of arable lands, it is not easy to determine if there was a core area or infield, except that the farmstead that was abandoned in the early nineteenth century resides at the north end of the site. However, this superseded the township on Stroan Hill in the middle of the site, where there are several houses and yards and a kiln, all of which went out of use before the rig. Thus not only has the farm seen a process of piecemeal enclosure that was still in progress when it was abandoned, but the rig has gone out of use in some areas and continued in use in others. There are also small plots of lazy-bed rig within the site, which are not immediately evident from the plan and may be seen at other townships in the area (e.g. Laughenghie nearby, as well as at High Eldrig in Wigtownshire, Pl. 2.11). The reason for this dichotomy in the method of cultivation is not understood, but it may be cultivation by cottagers of corners that could not be cultivated by a plough. However, more extensive patches of lazy-beds are to be found at Laughenghie and elsewhere in the south-west (e.g. Auchensoul, Ayrshire).

4. A forest stead at Martinlee Sike, Southdean, Roxburghshire (Fig. 2.6)

At Martinlee Sike, Southdean in the royal forest of Jedburgh in the Borders, 7ha are enclosed in a D-shaped area on the north bank of the Carter Burn, which defines the enclosure on the south. The enclosing dyke comprises a bank with external ditch. Such an arrangement is to be found in England enclosing forest assarts as far apart as Northumberland (Dixon 1983, RCAHMS 1994) and Dartmoor, where it is known as a 'corn-ditch' (Fleming and Ralph 1982). An extension to the west takes in another 14ha of which less than half is cultivated, possibly as a piece of outfield.

Pl. 2.13: Lazy-beds at Dunan, Traigh nan Sruban, Isle of Lewis.
Note the contiguous lazy-beds on the left of the photograph
(copyright CUCAP CCM 65).

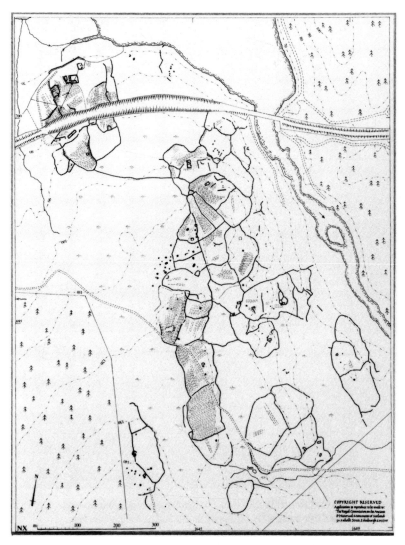

2.5: Plan of a narrow-curving rig field-system, Stroan, Kirkcudbrightshire.
(Copyright RCAHMS)

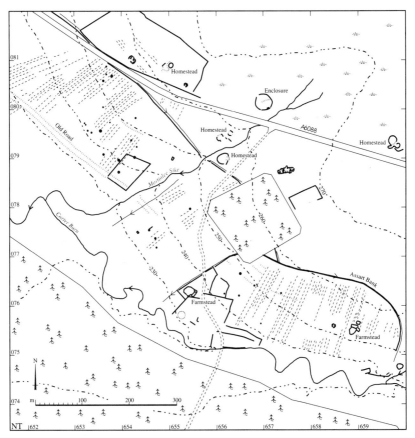

2.6: A plan of the forest stead at Martinlee Sike, Southdean, Roxburghshire
(copyright RCAHMS)

Elsewhere the same construction is used for the head-dykes of the rigged ground of forest steads at Northbank Tower, Slack's Tower and Watties Spindles. Whilst all these sites were inhabited until the eighteenth century, Martinlee Sike was abandoned prior to the mid-sixteenth century and does not appear in subsequent tax records or maps. Both of the farmsteads within the assart comprise an enclosed yard and a subtstantial building with out-houses. The rig is mainly broad, but also includes a small patch of quite narrow rig, whose affinities are not understood. However, there are also traces here of prehistoric cultivation-terraces and clearance heaps. At the west side of the D-shaped enclosure, there are several stock enclosures, probably belonging to a sheep-farming episode.

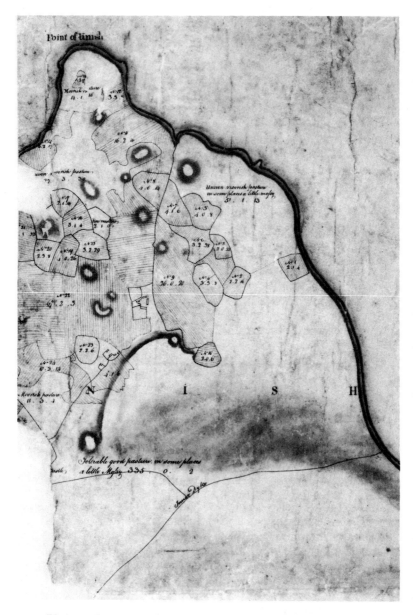

Pl. 2.14: Estate plan of Unish 1790 (copyright Macdonald, Stein, Skye).

Conclusion

I have tried to reveal something of the diversity of cultivation-remains in Scotland and of the distribution of its main types. Regional variation is self-evident. The form of rig in Galloway differs from that of the Cheviots, where there are special variants, whilst the north-west has a preponderance of lazy-bed rig, and forest areas may display special characteristics of enclosure with the establishment of medieval assarts.

Landscape and terrain are important formative considerations in Scotland and are particularly restrictive to agriculture in lazy-bed areas like Wester Ross or Assynt. However, this is also true to a lesser degree of central and eastern Scotland, where there were areas of moor and badly-drained bog up until the eighteenth century. The preference for lazy-beds on the north-west may be understandable in view of the limited arable potential of the soils. However, the plough has been used in the last century in these areas as at Unish on Skye. Lazy-bedding is more labour-intensive than ploughing, though more productive, an important consideration if arable is at a premium. The special differences of Galloway with its narrow curving rig and the occasional use of spade-dug rig, is difficult to explain, except as a reaction to local topography. Narrower curving rig, however, may prove more typical of other upland or Highland landscapes such as Perthshire, where it has recently been recorded in Strathbraan.

Questions of infield-outfield may be reducible to the division between the rig-covered ground, with or without a head-dyke, and the common waste beyond, although it is quite obviously more complex. Few areas of the country were without waste ground before the Improvements, but it is unlikely that arable remained static. The construction of a head-dyke might suggest a degree of stability in a field-system, but as I have suggested a head-dyke is not an essential type-fossil of pre-Improvement field-systems, and is often late in the sequence of the field-system. This does not preclude the taking in of patches of ground for occasional cultivation, as may be the case in the broken ground of the south-west (e.g. Stroan, Fig 2.5), or as is evident at some shieling grounds in Sutherland (e.g. RCAHMS 1993b), whilst there are the extensive areas of grooved broad rig in the Cheviots which may belong to phases of outfield cultivation.

Rig apart, the progress of enclosure appears to have been much more complex than has hitherto been considered. Piecemeal enclosure is apparent in many areas as well as enclosure for sheep, involving the complete eclipse of rig-systems. Whilst many enclosures are visibly late and part of the general process of agricultural improvement, some will be earlier, as in medieval assarts, if not earlier still, where pre-medieval fields have been re-used.

A great deal more mapping of complete field-systems is needed in many parts of Scotland if we are to understand the distribution of the different types of field-system or the variety of rig-types employed and why. Selective small-scale excavation of the range of field-systems is needed to provide a chronological framework for the relative sequence derived from field-survey. All the above will of course take time, but I expect that whatever I have written will be superseded if my suggestions are carried out.

Bibliography

Barrow G W S. 1962. Rural Settlement in Central and eastern Scotland: The Medieval Evidence. *Scottish Studies* 6: 123-144.

Bowen H C. 1961. *Ancient Fields: A tentative analysis of vanishing earthworks and landscapes.* London.

DES. 1987. *Discovery and Excavation in Scotland.* Edinburgh.

Dixon P J. 1983. Alnhamsheles. In: *Medieval Village Research Group 31st Annual Report*: 16.

Dodgshon R A. 1980. The origins of traditional field-systems. In: Parry M L, Slater T R (eds). *The Making of the Scottish Countryside.* London.

Edwards K J. 1980. Excavation and environmental archaeology of a small cairn associated with cultivation ridges in Aberdeenshire. *Proc Soc Antiq Scot* 109: 22-29.

Fairhurst H. 1968. Rosal - a deserted township in Strathnaver, Sutherland. *Proc Soc Antiq Scot* 100: 135-69.

Fenton A. 1963. Early and Traditional Cultivating Implements in Scotland. *Proc Soc Antiq Scot* 96: 264-317.

Fleming A, Ralph N. 1982. Medieval Settlement and Land Use on Holne Moor, Dartmoor: The Landscape Evidence. *Medieval Archaeology* 26: 101-137.

Graham A. 1939. Cultivation terraces in south-eastern Scotland. *Proc Soc Antiq Scot* 73: 289-315.

Guilbert J M. 1979. *Hunting and Hunting Reserves in Medieval Scotland*, John Donald, Edinburgh.

Guilbert J. 1983. The Monastic Record of a Border Landscape 1136-1236, *Scottish Geographical Magazine*, 99 Pt.1: 4-15.

Hall D N. 1982. *Medieval Fields.*Aylesbury.

Jirlow R, Whitaker I. 1957. The Plough in Scotland. *Scottish Studies* 1: 71-94.

Lerche G. 1986. Ridged fields and profiles of plough-furrows. *Tools and Tillage* 5.3: 131-56.

Parry M L. 1976. A typology of cultivation ridges in Southern Scotland. *Tools and Tillage* 3.1: 3-19.

Pollock D. 1985. The Lunan Valley Project: medieval rural settlement in Angus, *Proc Soc Antiq Scot* 115: 357-99.

Raine J. 1852. *The History and Antiquities of North Durham.*

RCAHMS. 1967. *Peeblesshire: An Inventory of the Ancient Monuments.* HMSO.

RCAHMS. 1990. *North-east Perth: an archaeological landscape.* HMSO.

RCAHMS. 1993a. *Waternish, Skye and Lochalsh District, Highland Region: An Archaeological Survey.*

RCAHMS. 1993b. *Strath of Kildonan: An Archaeological Survey.*

RCAHMS. 1994. *Southdean, Borders: An Archaeological Survey.*

Stat. Acct. 1791-9 *Statistical Account of Scotland.* Edinburgh.

Stevenson R B K. 1947. Farms and Fortifications in the King's Park, Edinburgh, *Proc Soc Antiq Scot* 81: 158-170.

Whittingdon G. 1973. Field Systems of Scotland. In: Baker A R H, Butlin RA (eds). *Studies of Field Systems in the British Isles.* Cambridge University Press, Cambridge: 530-79.

Wilson J. 1902. Farming in Aberdeenshire Ancient and Modern. *Transactions of the Highland Agricultural Society of Scotland*, 14, 5th Series: 76-102.

Chapter 3

Rethinking Highland Field Systems

Robert A Dodgshon

Runrig townships in the western Highlands and Islands are well-covered by late eighteenth and early nineteenth century surveys compiled prior to the spread of crofting and the clearances, but the lack of explicit documentary data for earlier periods means that our understanding of such townships and their field systems lacks historical depth. In the circumstances, establishing a deeper understanding necessarily depends on how well we exploit the archaeological potential of the problem. Discusion of this archaeological potential can be organised around three levels of approach. First, there is the problem of establishing what actually exists on the ground. Second, there is the task of sifting and sorting this data into categories of related features, establishing any pattern that might exist. Third, once we have established these patterns of related data, this can provide the basis for their interpretation as functional systems. Ideally, these interpretations should include ideas on how particular field patterns were exploited within a system of husbandry and how they were linked to a wider context of land tenure, landholding and settlement.

I have distinguished between these three levels of analysis and interpretation because each is a separate question and poses different problems. If we take them in turn, the first – establishing what data actually exists – might seem the most straightforward. However, in the context of the western Highlands and Islands, it is anything but straightforward. The problem has as much to do with what we take into the field as with what we find when we get there. I would suggest that prior assumptions have played a powerful role in constructing what we see in the region. These assumptions have to do with the antiquity or archaism of what existed prior to the radical changes wrought by the re-organisation of some farming townships into crofts and the clearances of others to make way for sheep farming or deer forests. Some – but by no means all – of these townships were farmed by small groups of tenants holding their land in the form of runrig open fields. At most, the open-field systems represented by these runrig townships rarely involved more than 100 acres of arable on the mainland though on Hebridean islands like Tiree or South Uist, they could extend to over 300 acres and, exceptionally, to over 500 acres of arable (Dodgshon 1993a). This arable was cultivated on either an arable-grass or an infield-outfield basis. The challenge for landscape studies in the western Highlands and Islands is that runrig and infield-outfield have been too easily portrayed as archaic institutions, with runrig being presented as the oldest type of open or sub-divided field system and infield-outfield as an early or primitive form of field cropping, even though no direct or convincing evidence has been published in support of such early origins. If guided by such assumptions, we would expect the pre-crofting and pre-clearance landscape to be a very old landscape, one adapted to the needs of such an open-field system over many centuries. With due allowances for fluctuations in population, this pre-crofting landscape might be seen as depicting not only the early modern landscape of the region, but also the medieval and, for some, may even be a window on the Iron age (to borrow Kenneth Jackson's phrase). If we assume that pre-crofting townships are built around what *ab origine* are archaic forms and institutions, then patently, it colours what we might expect to find on the ground for the medieval and early modern periods. We might expect signs of quantitative change – of growth and contraction – but not necessarily of systemic or qualitative change, that is, of fundamental changes in the nature of field layout either because of changes in tenure or landholding or because of change in the nature of the field economy. Indeed, as a relatively small open-field system, one subjected to over-development by the demands of crofting and sheep farming, there is no reason to suppose that runrig has left abundant traces on the ground. Furthermore, if we go into the field with the assumption that township layouts had been adjusted to the needs of runrig over many centuries, then neither can we expect any field layout that preceded runrig to have left abundant clues as to its character.

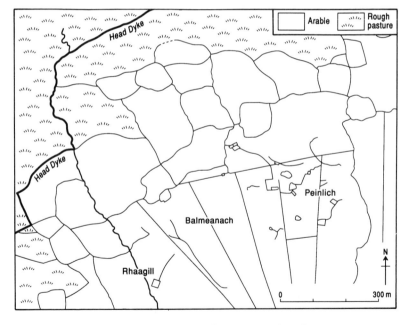

3.1: Surviving dykes and enclosures in the former townships of Rhaagill, Balmeanach and Peinlich (Glen Hinnisdal, Skye). Area delimited as arable based on Mathew Stobie's 1764 and 1766 plans, both of which cover Glen Hinnisdal, SRO, RHP 5992 and 5993. The early–mid nineteenth-century crofting layout liesacross the south-eastern corner of the map.

Plate 3.1: Cultivation rigs at Europie (Ness, Lewis).
The dyke running across the middle of the photograph represents the boundary dyke of the crofting system. As can be seen, cultivation rigs are present outside as well as inside it. In the top half of the photograph, cultivation rigs can be seen running obliquely across the line of the crofts despite the attempts by some crofters to remove them.

By way of a counter argument, I want to propose that these assumptions of archaism are ill-founded. When we focus on individual townships along the western seaboard or in the Hebrides and, where possble, remove the recent field layouts and surface disturbance created by crofting and sheep farming, we are left with more than the residual traces of a runrig system. Their pre-crofting or pre-clearance layout show signs of a more complex, compound character. The fact that there has been little debate over this field evidence suggests that the assumptions which we have used may have inhibited rather than aided our interpretation of how runrig has developed. Two basic types of evidence can be discerned: cultivation rigs, an obvious and widely appreciated form of evidence, and field dykes or enclosures, a less widely appreciated form of evidence.

Cultivation rigs

There are still abundant local traces of pre-crofting cultivation rigs. Logically, these are most easily detected where runrig townships were cleared for sheep or where the reorganisation of runrig townships into crofts involved a displacement of the township so that part of the pre-crofting layout was left undisturbed by the change. Yet even where crofts were laid out over the pre-crofting arable, the basic form of pre-crofting cultivation rigs might be left undisturbed. We can see both these possibilities at a site like Europie, Lewis, with well-preserved pre-crofting cultivation rigs surviving both outside the croft system and within it (e.g. OS NB520657). In some cases, those surviving within it run tangentially across the longitudinal line of crofts. The fact that some crofters have clearly tried to alter the alignment of their rigs so as to make them run in line with the croft merely serves to emphasise the inherited nature of those on either side (Plate 3.1).

Across the western Highlands and Islands, there are many townships in which pre-crofting and pre-clearance cultivation rigs have survived on a scale sufficient to enable us to reconstruct coherent patterns of layout. Their potential in helping us to elucidate the layout of pre-crofting or pre-clearance townships has been shown by work on Arnol, Lewis (Arnol, Department of Geography, UCW 1978), and the Vaternish Peninsula, Skye (Dixon 1993). Elsewhere, similar work on Rum enables us to see just how broken and opportunistic the arable of such townships could be on difficult, marginal sites (Rum, University of Glasgow 1972).

Beyond mapping such rigs, further pattern might be recovered by asking whether different types of rig reflect the use of different tools or implements like the spade, caschrom and plough. My own view is that there is no easy rule-of-thumb guide on this matter. At the extremes, narrow, sharply raised rigs are likely to be associated with the spade or caschrom whilst broad flat rigs are likely – though not exclusively – to be associated with the plough, but that still leaves many in between that cannot be so easily classified. In reply, it might be argued that these problems of rig classification are not really important. After all, are not spade and plough rigs part of the same system of cultivation? My own view is that we need to be more cautious over how we answer this question. The social and economic conditions under which hand tools like the spade and caschrom were used may have been quite different from those under which the plough was used. Yes, hand tools are more suitable for poor, broken or steep land, but they also require substantially more labour per acre than the plough, so much so, that it is unlikely that they would have been widely used except under conditions of land pressure when the combination of heavy local populations and limited quantities of arable per family made their use attractive given the extra yield they provided (Dodgshon 1992). In short, there is a case for arguing that the widespread use of the spade and caschrom do not necessarily manifest the survival of a deep-rooted cultural practice but historically-specific conditions of population pressure. Seen over time, some townships may even have switched between hand tool cultivation and plough cultivation as population growth or decline altered the balance between land and labour. In other words, if we can distinguish betwen them, there may be a case for seeing spade and plough rigs as two separate patterns rather than different parts of a single pattern. At Bragar, Lewis (OS NB292490), for instance, broad 'plough' rigs have been neatly cross-sectioned, furrow to furrow, by narrow rigs typical of the spade or caschrom in a way that makes it difficult to see them as part of the same pattern.

Runrig and its antecedents?

As a system based on sub-divided fields, one supposedly matured over many centuries, it would be reasonable to assume that the typical runrig township was an open-field system in the literal sense, without any internal enclosure of arable. In fact, when we try to reconstruct what existed prior to crofting and the clearances by removing the overlay of modern dykes and enclosures, we are left with a farming landscape that is still characterised by enclosures. Furthermore, if we compare the layout of these enclosures with available estate plans, they appear to sub-divide the arable of pre-crofting and pre-clearance townships. In other words, they represent a problem that has to be explained in relation to the history of such townships. Such a relationship is especially clear where the switch to a crofting layout involved a shift in the layout of arable or where a township site was abandoned entirely. Both these possibilities can be illustrated by the townships of Glen Hinnisdal (Skye). Altogether, the Five Penny Land of Glen Hinnisdal was divided into five townships, each of one pennyland. Their layout, including that of arable, is depicted by Mathew Stobie's plans of 1764 and 1766, Glen Hinnisdal being a part of Stobie's plans of both Kilmuir, 1764 (SRO, RHP 5992), and Uig, 1766 (SRO RHP 5993). Between two successive estate surveys compiled in 1811 and 1851, the three townships of Balmeanach, Peinaha (or Glentinistle) and Upper Glen (or Glenuachdarach) were re-organised into crofts, whilst the most westerly township, Rhaagill, was abandoned (Macdonald Papers, Armadale, GD221/116, Report on Lord Macdonald's Estate, 1811; Hull University Library, DDKG/103, Memorandum and Report on His Lordship's Property of Macdonald and Sleat, 1851). The fifth township, Peinlich, was reported as still held by runrig in 1851 but had been converted to crofts by the time of the first Ordnance survey in 1875 (OS 1st edition, 6 inch 1875). When the dykes that criss-cross these townships are mapped, they not only extend over the arable of Balmeanach, Peinlich, Peinaha and Upper Glen that was left out of their re-organisation into crofts but they also extend across Rhaagill making it absolutely clear that, as a field problem, they need to be seen in relation to the pre-crofting or pre-clearance landscape (Fig. 3.1). Likewise, when we look at a township like Bragar on the north-west coast of Lewis, the arable abandoned by the switch to crofting carries traces of dykes and enclosures as well as those parts of the new crofting layout that were laid out over the pre-crofting arable (Dodgshon 1993b).

The dykes associated with pre-crofting and pre-clearance townships vary widely in construction. Some are dry-stone dykes, others turf-and-stone and still others, earthen banks. Eighteenth-century bylaws for Skye laid down very clear guidelines over how dykes on the island were to be constructed. Earthen dykes were required to 'have two faces six feet high', whilst dykes 'faced with stone' were to be five feet high and 'dykes of dry stone to be four & one half feet' (SRO, GD403/40/1-2). The better-preserved turf-and-stone dykes still to be found in some Skye townships conform closely to the dimensions given here. Elsewhere, those present on Coll are a mixture of low turf-and-stone walls built to a height of about three feet and dry-stone walls of varying height. The pre-improvement dykes present on Tiree, meanwhile, comprised earthen dykes which, according to Dr. Walker's report on the island, were built to a height of 5-6 feet (McKay 1980).

In addition to these differences in construction, there are further differences stemming from the fact that dykes have survived down to the present day to varying degrees. In some townships, like the abandoned township of Borrafiach (OS NG2363) in Waternish, or Eyre (OS NG4153) and Kendram (OS NG4373) in Trotternish, they have survived in good condition (Plate 3.2). By comparison, in Bragar (Lewis), they have been heavily robbed, most now being reduced to wall footings or stone trails. It is possible that many dykes at Bragar were robbed of stone to help build the dense network of croft boundaries that now lie to the south-east or to build the increased number of croft dwellings that developed on the site over the nineteenth century. However, there are signs that some dykes on the pre-crofting arable were reduced to a line of small cairns so as to enable rigs to be pushed through them (OS NB291490).

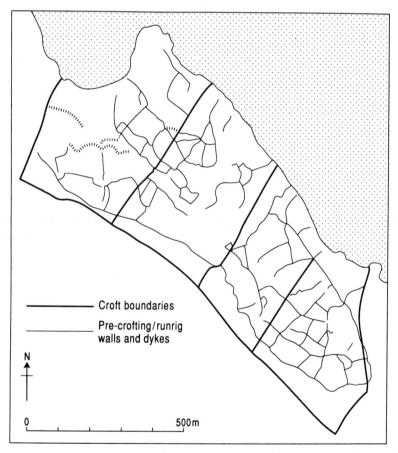

Croft boundaries

Pre-crofting / runrig
walls and dykes

N

0   500m

3.2: Surviving dykes and enclosures in the former township of Soriby, Ulva.
The croft layout is based on National Library of Scotland, *Plan of the Islands of Ulva,
Gometra, etc.* by John Leslie, 1812.

Plate 3.2: Enclosures on the former township of Kendram (Kilmuir, Trotternish, Skye).
Comparison with Mathew Stobie's 1764 plan of the area (SRO, RHP 5992) shows that
the enclosures sub-divide the area delimited as arable in 1764.

When we map what has survived in some townships, pre-crofting and pre-
clearance dykes form comprehensive and complex patterns of enclosure. Figure
3.2 shows the pattern of dykes and enclosures that survive in the former township of Soriby on Ulva. These form part of an extended pattern of enclosure associated with a string of small townships that run along the lower coastal slopes
on the north-eastern side of Ulva. Continuous patterns of enclosure can be found
elsewhere, such as in the south-western part of Coll. In most cases, though, we
are dealing with scattered blocks of enclosure rather than patterns that run continuously between townships.

Though we can map enclosures as if they formed one coherent pattern, there
is a case for arguing that most of what has survived from before crofting and the
clearances is composite, made up of different layers of development (cf. Armit
1989). This has a bearing on how we account for enclosures. Though an open-
field system, traditional runrig townships had need of some dykes and enclosures. First and foremost, there was a need for head dykes to separate grazings
from arable. Some eighteenth century sources refer to both an inner and outer
dyke (SRO, GD46/1/227; Bil 1990), one separating out arable from outfield,
meadows and winterings and the other separating winterings from hill ground
(cf. Rønneseth 1974; Widgren 1983). Of course, if arable was expanded, there
would have been a need for moving inner dykes and even the head-dyke, perhaps leaving earlier dykes as fossil forms. There are signs of the head-dyke at
Greaulin (OS NG395678) having been adjusted outwards both at its northern

and southern ends. There is also a case for arguing that the head-dyke at Borrafiach, Vaternish (OS NG235637) may have a complex history behind it. Amongst the other dykes that one might expect to find in a runrig system are those bordering cattlepaths. These enabled stock to be led from the common pasture across arable to byres and associated homesteads which were located in the centre of a township. On Skye, bylaws required that 'to all Farms or rooms of Land there should be a common path with a gate on the march dyke' (SRO, GD403/40/1-2). Not all townships, though, appear to have maintained such cattlepaths or their fences. A report on townships in Netherlorn in 1785 made the point that some tenants had 'their houses together with those of their servants all set down in one place, perhaps, in the middle of the farm, by which much mischief is done to their corns, by their cattle going to & from the pasture through unclosed fields' (SRO, GD112/12/1/2/14). Many tenants also maintained kailyards. These have survived even where townships have undergone reorganisation into crofts. Some of the large townships to be found along the north-west coast of Lewis (i.e. Bragar, Arnol) provide fine examples of kailyards dating from the pre-crofting period that now appear to lie haphazardly across the modern crofting layout. Where settlement was clustered, we can also expect some townships to have linked farmstead to farmstead, kailyard to kailyard, so as to create an enclosed space at the very core of the settlement for controlling stock. Hints of this are still apparent around the pre-crofting settlement at Peinlich, Glen Hinnisdal. Finally, bylaws for both Skye and Lewis make it clear that townships were also required to maintain a penfold for keeping stock under control at night.

In aggregate, this list of dykes and enclosures compatible with a runrig system is sufficient to explain a fair proportion of those to be found within pre-crofting and pre-clearance townships. Even so, I would argue that it is still not sufficient to explain the full range of dykes and boundaries actually present in such townships. In some cases, but especially in townships which have not experienced over-development, we are confronted with a fairly full sub-division of arable by enclosures. Even in townships which have experienced much recent change, enough remains to suggest that some must once have been enclosed to a degree one would not expect of an open-field system. Much basic research work still has to be done on these systems of enclosure but, as a working proposition, I want to suggest that they constitute a form of township organisation that needs to be distinguished from runrig, rather than confused with it; that is, they need to be explained in their own terms. There is nothing in the nature of runrig that provides an obvious raison d'etre for them.

Considered hypothetically, the presence of enclosures within runrig townships can be interpreted in at least four ways. First, there is the possibility that they simply represent consumption dykes, an attempt to clear arable of excess stones. On particularly stony soils, such as one has in many parts of the western Highlands and Islands, dykes would have been a better solution than clearance cairns since there was a need to clear large amounts of stone. The need to clear fields would have been present whether fields were held in runrig or not, with the benefits of having a stone-free surface presumably outweighing the diseconomies of having an enclosed 'open'-field system. There is, of course, no way of testing for this interpretation, but the fact that many townships also cleared stones to the furrows that lay between their rigs could mean either that building dykes around fields was not a sufficient solution to this problem or that dykes had other purposes. The extensive use of turf or feal for dyke building would also suggest that clearing the arable of stones was not the only reason for dyke construction.

A second explanation might be to see enclosures as a by-product of how land was colonised. If each field was colonised individually, then a bounding dyke might initially have been needed to protect its arable from encroachment by stock at least until communities built a general-purpose head-dyke. Local variations in land quality or ground conditions may have encouraged some communities to colonise scattered pockets of potential arable, enclosing each separately from the surrounding waste or pasture. Only, later, as a secondary stage, might such fields have been linked to form a continuous pattern of enclosures. If we look at fields morphologically, there is no doubt that some enclosures must have been colonised as detached or freestanding units, but this cannot serve as a covering explanation for all enclosures. Indeed, if farming was organised around runrig communities, then there is just as strong a case for supposing that townships would have built a general- purpose head-dyke or ring-garth at an early stage, thereby dispensing with the need for enclosing arable, parcel by parcel (cf. Ronneseth 1974).

A third possiblility is to see enclosures as making a fairly late appearance in the landscape by linking them to the greater interest in the marketing of cattle that spread through parts of the region over the late seventeenth and early eighteenth centuries. Again, the problem with such an interpretation is that it cannot be applied generally. A small number of farms in one or two areas (i.e. parts of mainland Argyll), particularly those held by the wealthier tacksmen, had certainly shifted into specialised cattle production on a scale that might have encouraged investment in enclosure by the early eighteenth century, but elsewhere, the increased marketing of cattle was associated with runrig townships and reflected changes in the composition of rent rather than any fundamental change in the organisation of husbandry. As estates commuted payments in kind into cash rents over the late seventeenth and early eighteenth century, many tenants took the opportunity to shift the rent burden from arable to stock by selling one or two cattle annually so as to raise the increased cash now needed for rent. This adjustment, though widespread, is unlikely, in itself, to have fostered the building of enclosures across arable. Besides, the layout of enclosures in terms of their size and shape, coupled with the fact that some townships appear to have robbed them freely for building stone, is simply not compatible with an interpretation that portrays them as a late seventeenth or early eighteenth century 'improvement'.

A possible variant of this last interpretation might be to link enclosures to the fact that not all farms were set in tenantry or farmed as runrig townships. Some were farmed in hand by the tacksmen who held the lease. Even before the seventeenth century and the expanded market for cattle, we can expect these farms to have had a greater orientation to livestock, not least because of the role played by livestock – especially cattle – in the building of social status through feasting, payments of *tocher gude*, fostering deals and so on. For this reason, they may have had need of some enclosures long before the growth of any specialised interest in the marketing of cattle. But again, as a general explanation for the dykes and enclosures present in the western Highlands and Islands, such an interpretation is weakened by the fact that dykes and enclosures occur in former runrig townships and not just on those farms held and farmed *in toto* by tacksmen without any sub-letting.

Each of the interpretations considered so far would still leave us with a medieval landscape made up of runrig townships. By comparison, the fourth explanation requires a radical rethink of what may have existed. In practice, any sort of enclosure, other than head-dykes, cattlepaths and kailyards, inhibits the workings of an open-field system, particularly the exercise of rights of grazings across the arable after harvest. Arguably, the amount of enclosure actually present in some townships goes beyond what would be expected within a mature open-field system. Rather than explain this inconsistency away by special pleading over the need for consumption dykes or how land was colonised, the most logical interpretation would be to see them as representing the surviving traces of a wholly different system of farming and landholding, one which preceded runrig. Such an interpretation is given support by four kinds of evidence. First, traces of enclosure are found across a wide range of townships, mainland and Hebridean, and in different kinds of area and location. In these circumstances, their existence requires a general covering explanation that addresses the institutional foundations of field layout either in terms of landholding, tenure or husbandry. Second, though some runrig townships appear to have left their dyke systems in tact, others have systematically reduced them or robbed them of stone. Such a response is best explained by seeing runrig as post-dating enclosures not vice versa. Third, cultivation rigs in some townships (i.e. Bragar) appear to have been constructed across or through dykes, establishing a demonstrable order of sequence. Fourth, the size and shape of some patterns of enclosure makes it difficult to see how they could have been intentionally laid out as the 'shotts' or 'furlongs' of a runrig layout. We can best rationalise these examples by interpreting them as inherited from an earlier system of landholding or husbandry and as incorporated into a runrig system only at a later stage. The function of enclosures within this earlier system may have been to differentiate the fields (as opposed to the strips) held by different landholders and/or to identify and control the access of stock to blocks of meadow or winterings. Certainly, it would be easy to envisage how particular patterns of enclosure, such as those skirting Lon Bhearnus on Ulva, might initially have comprised a cluster of arable enclosures on the better land of the coast, with those occupying the small shelves of land that step the slopes behind serving to enclose small pockets of meadow land or better pasture. In time, with the shift into runrig suggested here, together with population growth, arable and the runrig layout of holdings would have been extended across most enclosures without distinction. There may be an even broader perspective to this problem. Enclosures might be linked to the more dispersed pattern of settlement that is evident in some townships (Dodgshon 1994, *in press*). If so, then it may enable us to draw a broad contrast between, on the one hand, a landscape based on enclosures and more dispersed settlement and, on the other, one based on runrig and more nucleated settlement.

Establishing when such a shift might have taken place will not be easy to answer because the question at issue is not about when enclosures were built but the more difficult problem of when were they abandoned as a functioning system. Though very limited in what it can tell us about the problem, documentary evidence may be helpful in establishing the broad context under which such a switch may have taken place. Using such evidence for Lowland Britain, a case has already been made out for seeing open fields as having developed within a framework of feudal landholding (Dodgshon 1980). This is not to deny the contributory role of other factors such as population growth, risk aversion or scarcity of pasture, but such factors can only influence the problem if their solutions are seen as ring-fenced by the definition and boundedness imposed on landholding by feudal assessments. If the feudalisation of landholding is seen as a formative factor in creating runrig then it may provide a broad chronology for its development in the western Highlands and Islands. This is because landholding in the region is unlikely to have been feudalised until after the Treaty of Perth (1266) and even then, its feudalisation must have been a weak, slow process. In other words, we may have no reason to expect runrig to have developed until after the mid-thirteenth century. Of course, this represents only a *terminus post quem*. In some cases, the switch may have occurred much later. Clearly, this is a fundamentally different chronology for runrig's development to that conventionally assumed. Yet as an interpretation, it would position the debate over its origins more firmly within the debate over open fields generally. Recent work in lowland Britain has reached the conclusion that large, co-ordinated open-field systems did not come into being until the tenth–eleventh centuries (Taylor 1983; Dyer 1990). In parts of northern England, an even later date has been proposed. To suggest that runrig developed in the western Highlands and Islands only from the mid-thirteenth century onwards would be wholly consistent with this revised chronology. Indeed, even without being prompted by the field evidence reviewed above, the case for runrig's early origins would surely need to be revised in the light of this wider debate.

Such a re-interpretation clearly affects how we read the long-term history of west Highland and Hebridean landscapes. So long as runrig is seen as an archaic institution, then we would not expect such landscapes to reveal evidence of fundamental change. We might expect signs of quantitative change, that is, of townships expanding or contracting, and evidence for new sites being occupied or abandoned, but not signs of qualitative or systemic change, with shifts in the way farming, landholding and settlement were actually organised. If we accept the evidence for an enclosed landscape beneath runrig, then it clearly opens up a different perspective. The landscapes that existed in the region for much of the medieval period may have been quite different from those which prevailed during the early modern period. As a matter of urgency, we need to research the character of these enclosed landscapes and their functional and chronological relationship to runrig. If the patterns of enclosure evident in many townships do represent traces of a landscape that preceded runrig, then it offers considerable potential for reconstructing how west Highland and Hebridean landscapes may have been changed over time. To fully exploit this potential, we need to establish the institutional context of enclosed landscapes. Did they involve a different form of land tenure and landholding to that which we associate with runrig? How were they farmed? Given how little we know about social and economic change in the western Highlands and Islands over the medieval period, there may be important general insights to be generated here.

As regards what replaced such landscapes, or runrig, what I have proposed clearly downplays its role in the history of west Highland and Hebridean landscapes. It would now need to be seen as a phase of township organisation that prevailed only during the closing years of the medieval period and over the early modern period. Further work needs to address two key problems of field interpretation. First, I do not rule out the possibility that the runrig layouts of some townships, especially some of the larger Hebridean townships, may have had a degree of regularity about it, hints of which may be recoverable through the metrological analysis of surviving broad-rig forms. The second basic problem of interpretation concerns how much we can read from the field evidence concerning the resource strategies employed by townships. For example, important differences of organisation may be associated with the differences between what I would call a dung-based system as opposed to a seaweed-based system. This is a distinction which I have elaborated on in a paper published elsewhere (Dodgshon 1993a). Dung-based systems are likely to have been infield-outfield systems. Their more elaborate cropping structure may have involved the use of an inner as well as an outer or head dyke. Likewise, we can expect farmsteads to have had byres for the indoor wintering of stock and, if these farmsteads were located within the arable, we can expect to find cattlepaths leading stock from the common pasture across the arable, though its possible that a laithe-barn solution might have been adopted in some townships with byres scattered across the fields. To help fodder stock, we would expect townships to have maintained small areas of meadows, either for grazing or for the cutting of hay. By comparison, seaweed-based systems were largely based on grass-arable rotations. For this reason, there may only have been a single head-dyke. Also, given the tendency for seaweed-based townships to out-winter stock, there is no reason to expect cattlepaths, byres or hay-meadows to have been present. Indeed, landscape-wise, such systems may have left few obvious clues.

Bibliography

Armit I. 1989. *The Loch Olabhat Project North Uist 1989*. 4th Interim Report, Department of Archaeology, University of Edinburgh, Project Paper no. 12.

Arnol. 1978. Isle of Lewis, 1 sheet. Map of cultivation rigs and crofts prepared in the Department of Geography, University College of Wales, Aberystwyth.

Bil A. 1990. *The Shieling 1600-1840. The Case of the Central Scottish Highlands*. J Donald, Edinburgh.

Dixon P. 1993. A review of the archaeology of rural medieval and post-medieval settlement in Highland Scotland. In: Hingley R (ed). *Medieval or Later Rural Settlement in Scotland*. Historic Scotland Ancient Monuments Division, Occasional Paper no. 1, Edinburgh.

Dodgshon R A. 1980. *The Origin of British Field Systems*. Academic Press, London.

Dodgshon R A. 1992. Farming practice in the western Highlands and Islands before crofting: a study in cultural inertia or opportunity costs? *Rural History*, 3: 173-89.

Dodgshon R A. 1993a. Strategies of farming in the western Highlands and Islands of Scotland prior to crofting and the clearances. *Economic History Review*, XLV1: 679-701.

Dodgshon R A. 1993b. West Highland and Hebridean landscapes: have they a history without runrig? *Jnl. of Historical Geography*, 19: 383-98.

Dodgshon R A. 1994 *in press*. West Highland and Hebridean settlement: a study in stability or change? *Proc Soc Antiq Scot*, 123.

Dyer C. 1990. The past, present and future in medieval rural history. *Rural History*, 1: 38-40.

McKay M M (ed.). 1980. *The Rev. Dr. John Walker's Report on the Hebrides of 1764 and 1771*. J Donald, Edinburgh.

Rhum 1972. 4 sheets, 1:10000, map published by the University of Glasgow.

Rønneseth O. 1972. Gard und Einfriedigung. Entwicklungsphasen der Agrarlandschaft Jaerens. *Geografiska Annaler*, series B, special issue no. 2 (Medd. B29).

Taylor C. 1983. *Village and Farmstead*. London.

Widgren M. 1983. *Settlement and Farming Systems in the Early Iron Age. A Study of Fossil Agrarian Landscapes in Ostergotland, Sweden*. Stockholm Studies in Human Geography, no. 3, Stockholm.

Chapter 4

Soils and Landscape History: Case Studies from the Northern Isles of Scotland

Donald A Davidson and Ian A Simpson

Introduction

It is widely accepted by archaeologists that the investigation of soils on sites being excavated can make important contributions to the overall interpretation of former processes, functions and environmental conditions. Thus on-site soils are considered as integral to the archaeological record and contain evidence of cultural history. Unfortunately the same approach is not so evident when concern turns to landscape history. Soils tend to be viewed as providing the surface over which landscape evolution takes places. The aim of this chapter is to challenge this view by emphasising that landscape history is imprinted in the nature and properties of soils. Thus it will be argued that soils are integral to landscapes and merit investigation if the overall objective is a better understanding of landscape evolution.

Soils are dynamic and are thus subject to continual change. As soon as plants colonise a bare surface, the underlying material is modified as a result of soil forming processes which include the incorporation and decay of organic matter in soil, the effects of soil organisms, the movement of soil water with associated solutes and fine sediments and the weathering of rock fragments and minerals by physical, chemical and biological processes. It is the combination of such processes which results in the distinctive form or morphology of soils which is expressed in horizon sequences. Soils can be visualised as systems which owe their attributes to present as well as past processes including that of human activity.

In pedology it is widely accepted that soils are part of landscapes, with soil properties exhibiting spatial and temporal variability. Theories of soil–landscape systems have evolved over the past few decades, moving from two dimensional concepts, embodied in the concept of the catena (Milne 1935, Morison et al 1948) and the nine unit land surface model of Dalrymple et al (1968), through to three dimensional soil–landscape systems advanced by a number of workers (Huggett 1975; Vreeken 1973) which use the drainage basin as the basic functional unit. This chapter extends the concepts of soil–landscape systems by considering soils as integral to cultural landscapes.

The production of food by arable cultivation has an inevitable impact on soils. Vegetation is removed, soil is mixed, only a few plants are grown, nutrients are added and soil conditions are altered through drainage and stone removal. The impact of cultivation on soils becomes particularly marked in the more marginal areas since greater effort is required to sustain arable production. Two key processes are thus cultivation and manuring. This chapter demonstrates the impact of such processes on soil properties and then illustrates how soil investigation can contribute to an understanding of landscape evolution.

Three contrasting cultural landscapes with different chronologies from the Northern Isles are considered on a chronological basis to demonstrate this potential. The small island of Papa Stour on Shetland is first selected to illustrate the interdependence of soil and landscape history with particular reference to the nineteenth and twentieth centuries. Impacts during the Medieval and prehistoric periods are then discussed respectively for Marwick in West Mainland Orkney and Tofts Ness on Sanday, Orkney.

Papa Stour, Shetland

The small island of Papa Stour, which is located to the immediate west of Mainland Shetland, provides a splendid microcosm of a landscape system. The island, though suffering economically today from its peripheral location in western Shetland, was once much more prosperous. During Norse times the island was on the main 'fairway' and was thus of particular importance to Norse seafarers (Crawford 1984). Detailed historical and archaeological research by Crawford (1985; 1990) has identified the area known as 'Biggings' as the central area of settlement in the Middle Ages and thus the most likely site of the original Norse settlement (Fig. 4.1). Another cluster of farms was in the area known as East Bigging.

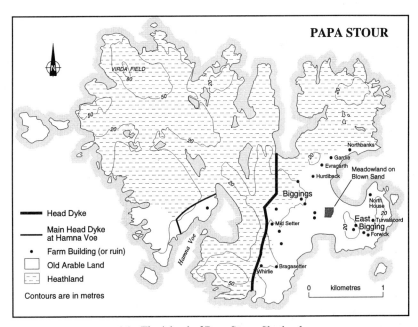

4.1: The island of Papa Stour, Shetland

To understand the present nature and distribution of soils on Papa Stour, reference has to be made to the land management systems of the nineteenth and early twentieth centuries. Fortunately a great deal of historical work has been done for the island by Fenton (1978) who describes in detail the results of a survey in 1846 by Thomas Irvine. The island consisted of one townland under one head dyke with only two outsets at Hamna Voe and Northbanks (Fig. 4.1). During the summer this dyke was the division between the townland to the east and the common grazings or scattald to the west. After the crops were harvested in the autumn, the stock were free to roam over the townland (Fenton 1978). The common grazings were an important source for fuel and winter bedding material for cattle. Peat was cut from the scattald for centuries, but had become extremely scarce towards the end of the nineteenth century. Fenton (1978) comments that many of the North House tenants had left by 1881 and new tenants from the Mainland were unwilling to come because of the lack of peat. Stripping of the shallow peaty layer in the west of Papa Stour also caused mineral material to be removed. The peat turf, after drying, was burnt in domestic hearths, the ash heaped in the middens and then spread on the surrounding fields. The peaty material with associated mineral matter was also used for winter bedding of cattle in order to soak up the animal wastes. Again, such material was ultimately added to the surrounding fields along with applications of seaweed. It is difficult to determine the relative proportion of hill peat cut for fuel rather than for byre bedding material. The resultant ash would have been added to midden heaps and later to the fields. However, ash from peat and turf fires was also used to soak up liquid waste in byres (Fenton 1978). The net effect of spreading midden material was the gradual deepening of topsoils. Such topsoils today are distinguished by their depth, often up to c80cm thick, and their apparent homogeneity. Over the centuries of topsoil formation, considerable human effort was thus spent on the maintenance and indeed enhancement of soil quality. This was further emphasised by human teams using delling spades in fields as small as 15m by 20m. The overall effect of the paring and burning of peaty topsoil was the degradation of the soil resource in the scattald to the west of the head dyke and the corresponding improvement in soil within the townland. The head dyke thus forms a fundamental division in the present day soil geography of Papa Stour.

West Mainland, Orkney

In Orkney the Bilbster soil series is a freely or imperfectly drained podzol developed on drift derived from the Stromness and Rousay Flags of the Middle Old Red Sandstone. A deep top phase within this series has been identified by the Soil Survey for Scotland (1981) which is confined to the West Mainland of Orkney and to Stronsay. Preliminary analysis of the deep top phase established top soil thicknesses of up to 115cm, increasing with proximity to early farmsteads, together with enhanced total soil phosphorus levels (93-1148mg/100g) of between two and three times the levels found in top soils of the parent Bilbster series. Both these characteristics are indicative of 'plaggen' soil formation and so the deep top phase of the Bilbster series can be regarded as an anthropogenic soil (Davidson & Simpson 1984; Simpson 1985). Plaggen soils resulted from the stripping of heather or grass turves which were then used as animal bedding in the byre. When the byres were cleaned out, the turf and manure mix was applied to the arable land where the mineral component attached to the turf gradually contributed to the creation of an artificial soil horizon.

The chronology of deep top soil formation has been established by radiocarbon dating and by association with cultural features of known age (Simpson, 1993). These approaches have dated the origin of deep top soil formation to the late twelfth/early thirteenth centuries AD, a period when temperatures were as much as one degree Celsius above what they are today and when, based on place name evidence, there was increased organisation of agricultural activity. Deep top soil formation continued for some 700 years until the late nineteenth/early twentieth centuries AD when new, inorganic, soil fertilisers became available to replace the less convenient and labour intensive plaggen manuring process. Relict properties of deep top soils may therefore permit reconstruction of cultural landscape processes over the late Norse and Medieval period of Orkney.

Deep top soils found in the Medieval coastal township of Marwick in West Mainland Orkney have been characterised using a number of properties to elucidate the processes of manuring and cultivation. These properties include particle size distribution, used to identify sources of inorganic material for deep top soil formation; stable carbon isotope ratios which differentiate between marine and terrestrial sources of carbon as well as making distinctions within terrestrial sources (Simpson 1985b); phosphorus chemistry which has the potential to identify intensities of manuring practice and associated land use functions (Eidt 1984) and thin section micromorphology, which permits the investigation of the fabric and organisation of the soil and will reflect, in part, soil organisation resulting from the process of cultivation. Comparative analysis of these deep top soil properties with the properties of other soils and sediments in Marwick allow conclusions to be drawn on manuring and cultivation processes.

Mean particle size distribution range of the deep top soils in Marwick is between 6.50-7.16μm. Comparing this range with particle size distributions of soils and sediments elsewhere in Marwick suggests that the late Norse and medieval farmer in this area of Orkney ignored a number of potential sources of inorganic material. The dominantly non calcareous beach sands were ignored, perhaps because of their lack of moisture retention and low nutrient value. Also ignored were gley soils from within the township which were presumably too wet, and podzols which formed part of the arable rig land subject to periodic redistribution among the farmers. It is difficult to obtain clear evidence of peat ash as a major input to the deep top soils. The inorganic materials of deep top soil formation in Marwick comprise exclusively of podzolic soil material derived from the common hill land which lay beyond the turf dyke delineating the extent of the township (Fig. 4.2). An estimated 199,300m³ of material has been moved from the hill land to the cultivated areas of Marwick mainly in the form of stripped grassy turves during the period of deep top soil formation. A potential contrast thus emerges between Papa Stour and Marwick in terms of the initial use of the stripped shallow peat or turf. For Marwick, bedding material for cattle was of paramount importance whilst this not so clear for Papa Stour.

The evidence of $\delta^{13}C$ values from the deep top soils of Marwick suggests that there is some diversity in organic formation materials. These values are within the range of -26.7 to -29.8 $^0/_{00}$ indicating organic amendments were dominantly terrestrial in origin with a minor marine seaweed component. It is also possible to discern distinctions within the terrestrial class of materials with the more negative $\delta^{13}C$ values reflecting a greater input of animal manure relative to plant materials. On the basis of these characteristics, animal manure application was greatest on the Muce and Netherskaill deep top soils while at West Howe, organic sources were more variable, involving a greater proportion of plant material (turf) relative to manure and a small amount of seaweed.

From the analysis of inorganic and organic sources it is clear that there was the import of material into the township of Marwick in order to sustain arable soil quality. This was dominantly in the form of turf from the hill land, but also involved seaweeds from the seashore. Use of animal manure represents the cycling of materials within the township rather than imports with the manure coming from livestock that were wintered in byres and fed on hay gathered from meadowland (Fig. 4.2).

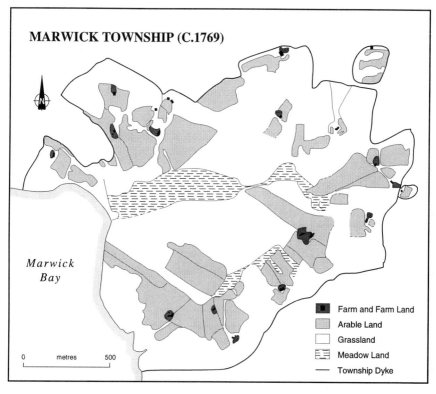

4.2: Land use in Marwick township, Orkney, 1769

Patterns of total phosphate within the deep top soils of Marwick and the surrounding soils allow inferences to be made about the organisation of cultivation practices within the landscape. Total phosphate levels from a transect at West Howe demonstrate a clear interface between the deep top soil edge and the other early arable land. This permits the inference that the deep top soil formed a disinctive part of the arable land area known as the *tounmal*, the land held by a farm semi-permanently and kept in continuous arable cultivation by a single occupier. Arable land beyond the area of deep top soil formed the *tounland* which was subject to periodic redistribution amongst the farms of the township and liable to being taken out of arable production on occasion. An important difference needs to be noted between tounmals in Orkney and Shetland (Fenton, 1978). In Orkney, such land was the best and was heavily manured to support the growth of bere. In contrast, tounmal on Shetland was an area of grazing or rough land which was never cultivated. On Papa Stour, all topsoil on the townland seems to have been deepened with the exception of grazing areas. Within the deep top soil areas of Marwick the general trend of total phosphate is for an increase with proximity to the farmstead and to the soil surface. On the basis of the link between total phosphate levels and cultivation activity it is clear that the intensity of cultivation increased over the period of deep top soil development and was of greatest intensity closest to the farmstead. Support for this interpretation is provided by the evidence in soil thin section of textural pedofeatures resulting from slaking effects associated with cultivation activity. Textural pedofeatures observed from an early phase of deep top soil formation are seen to be finer in texture than those from later in the development of the deep top soil. This implies that there was a shift from lighter to heavier cultivation implements as the deep top soil developed, reflecting technological change and the need to increase arable land productivity.

Tofts Ness, Sanday Orkney

One of the best surviving concentrations of prehistoric monuments in Orkney is to be found on the Tofts Ness peninsula located at the northern end of Sanday. The monuments in this archaeological landscape, together with their associated soils, date from the Neolithic to the Iron Age and have survived because of the peninsula's marginal aspect and by having been buried by extensive deposits of calcareous wind blown sand. A Neolithic/early Bronze age landscape has been fossilised by a period of sand deposition which formed the land surface of the late Bronze age/early Iron age landscape. This second landscape was in turn fossilised by a further period of major sand deposition forming the present day land surface (Dockrill et al, *in press*). The fossil landscapes of Tofts Ness provide an outstanding opportunity to examine early landscape processes and in particular, the evidence for manuring and cultivation of fossil soils. Investigations of the fossil soils are at an early stage but preliminary analyses have been undertaken using thin section micromorphology (Dockrill et al, *in press*).

Micromorphological evidence in the form of stone rims from the Neolithic/early Bronze Age landscape suggests that manuring and cultivation activity took place when processes of podzolisation were becoming more dominant. Early Neolithic cultivation may have contributed to these processes as in some parts of this landscape there is evidence of dusty textural pedofeatures which are frequently associated with cultivation activity. Indeed, in view of the small amounts of fine charcoal material, smooth but fractured phytoliths and amorphous organomineral fragments that are distinctly red in oblique incident light when observed in thin section, it is entirely probable that slash–burn was the first type of agricultural activity practiced in this area.

The podzolisation process together with the land management systems just described would have resulted in significant nutrient depletion and attempts to mitigate this trend during the late Neolithic/early Bronze age involved the use of organic manures and composts. Interpretation of thin sections from the Neolithic contexts clearly demonstrate this process with agricultural soils having marked amounts of aging arthropod faecal material, enhanced levels of highly decomposed amorphous plant remains and the occurrence of intra-aggregate channels, all of which are indicative of significant levels of biological activity associated with substantial manuring of soils. Examination of the Neolithic midden from the Tofts Ness study area gives some indication of the materials that may have been deposited on the arable land. Calcium oxalate spherulites and a few associated fragmented phytoliths indicate faecal material of herbivorous domestic animals as one likely source of material. Red fine material with embedded quartz grains observed in oblique incident light is indicative of turf ash material as a second possible source. Irrespective of the source of materials however, it is clear that attempts were made during the Neolithic period to maintain, or indeed to improve soil quality.

During the late Bronze and early Iron age, wind blown calcareous sands that covered the Neolithic landscape created an environment that was highly marginal for agriculture. Sandy textured soils are susceptible to wind erosion and are droughty for arable crops during the growing season. To enable cultivation of these soils manures were used to stabilise the soil, increase soil water holding capacity during the growing season and improve soil nutrient levels. Analysis of one soil from this fossil landscape using thin section micromorphology provided evidence of ash and manures being used for this purpose with enhanced faecal material. A second soil from this landscape also demonstrates attempts to manage this marginal environment for agricultural purposes. The context of this soil, sandwiched between bands of calcareous sand clearly indicated that it was anthropogenic in origin, purposively created by the importing of inorganic and organic materials from another part of the landscape, a process clearly comparable to plaggen manuring. The quartz content of this soil suggested turves had been used together with midden material, evidenced by burnt bone fragments among other materials. Silty clay infills observed in thin section testify to the intensively cultivated nature of this soil. Clearly the use of manures in this early cultural landscape turned an environment that was unviable for arable activity into one that was marginally viable.

Discussion

These man-made or anthropogenic soils in Papa Stour, the West Mainland of Orkney and Tofts Ness need to be viewed in a wider European perspective. Such soils, up to a metre or more in thickness and characterised by enhanced phosphorus levels and dating from at least the Iron Age, are found in many areas of north west Europe. Such soils have resulted from the transport of considerable volumes of material to the cultivated, arable areas of the landscape in an effort to maintain or enhance soil quality. Detailed research into the formation processes of anthropogenic soils has focused on the plaggen soils occurring on the Pleistocene sands of Belgium, Germany, Denmark and the Netherlands (Pape 1970; de Bakker 1980; van de Westeringh 1988; Mucher et al 1990).

Similar soils to the continental 'plaggen' soils have been identified in a number of places in Scotland other than Orkney and Shetland. Unfortunately there has been little pedological and historical research on these soils. Early work was by Glentworth (1944) who identified what he considered to be anthropogenic soils of between 30cm and 75cm thick in the Insch basin, Aberdeenshire. Barber (1981) found anthropogenic soils on Iona, formed between the seventh and eleventh centuries AD during a major period of monastic activity; another monastic connection to anthropogenic soil formation in Scotland is proposed by Romans (personal communication and cited by Barber 1981) for anthropogenic soils found around Fearn Abbey in Easter Ross.

The plaggen soils of Papa Stour and Orkney are human artifacts and thus are integral elements in the cultural landscapes of these islands. As outlined in this chapter, the scientific investigation of such soils can yield additional information on the nature of land management systems in the past. Such plaggen soils are the clearest expression of human impact on soils in Scotland, but it should be noted that there are many other instances of substantial soil change arising from past management practices. Obvious examples are stone clearance which is particularly outstanding in north east Scotland, the installation of land drainage schemes and the reclamation of low lying areas such as the carselands of the Forth or Tay.

In conclusion, an appreciation of soils as cultural elements in landscapes has three important implications. Soils are artifacts of human activity and ought to be routinely recorded along with other forms of environmental or structural evidence. Soils have histories which can be unravelled and thus assist with the broader interpretation of landscape evolution. The recognition of soils as artefacts has important implications in terms of developing conservation policies designed to assist with the management of archaeological landscapes.

Bibliography

Bakker de H. 1980. Anthropogenic soils in the Netherlands. *Rocznik Gleboznawcze* 21, 323-28.
Barber J W. 1981. Excavations on Iona, 1979. *Proc Society Antiq Scot* 111, 282-380.
Crawford B E. 1984. Papa Stour: survival, continuity and change in one Shetland island. In: A. Fenton and H. Palsson (eds). *The Northern and Western Isles in the Viking World,* John Donald, Edinburgh

Crawford B E. 1985. The Biggings, Papa Stour. In: B. Smith (ed) *Shetland Archaeology*. The Shetland Times, Lerwick

Crawford B E. 1990. Excavations at the Biggings, Papa Stour, Shetland. *Acta Archaeologica* 61: 36-43.

Dalrymple J B, Blong R J, Conacher A J. 1968. A hypothetical nine unit landsurface model. *Zeitschrift fur Geomorphologie* 12: 60-76.

Davidson D A, Simpson I A. 1984. Deep top soil formation in Orkney. *Earth Surface Processes and Landforms* 9: 61-81.

Dockrill S J, Bond J M, Milles A, Simpson I A, Ambers J. (in press). Tofts Ness, Sanday, Orkney: an integrated study of a buried Orcadian landscape. In: Luff R M, Rowly-Conwy P A (eds) *Whither Environmental Archaeology?* Oxbow, Oxford

Eidt R C. 1984. *Advances in Abandoned Settlement Analysis.* University of Wisconsin, Milwakee.

Fenton A. 1978. *The Northern Isles: Orkney and Shetland*. John Donald, Edinburgh.

Glentworth R. 1944. Studies on the soils developed on basic igneous rocks in central Aberdeenshire. *Transactions of the Royal Society of Edinburgh* 61: 155-56

Huggett R J. 1975. Soil landscape systems: a model of soil genesis. *Geoderma* 13: 1-22

Milne G. 1935. Composite units for the mapping of soil associations. *Transactions of the Third International Congress of Soil Science* 1: 345-47

Morison C G T, Hoyle A C, Hope-Smith J F. 1948. Tropical soil-vegetation catenas and mosaics. A study in the south-western part of the Anglo-Egyptian Sudan. *Journal of Ecology* 36: 1-84.

Mucher H J, Slotboom R T, Ten Veen W J. 1990. Palynology and micromorphology of a man made soil. A reconstruction of the agricultural history since late medieval times of the Posteles in the Netherlands. *Catena* 17: 55-67

Pape J C. 1970. Plaggen soils in the Netherlands. *Geoderma* 4: 229-56

Simpson I A. 1985a. *Anthropogenic Sedimentation in Orkney: The Formation of Deep Top Soils and Farm Mounds.* Unpublished PhD Thesis, University of Strathclyde, Glasgow

Simpson I A. 1985b. Stable carbon isotope analysis of anthropogenic soils and sediments in Orkney. In: Feiller N R J, Gilbertson D D, and Ralph N G A (eds) *Palaeo-environmental Investigations: Research Design, Methods and Data Analysis*. British Archaeological Reports (International Series) 258, Oxford

Simpson I A. 1993. The chronology of anthropogenic soils in Orkney. *Scottish Geographical Magazine* 109: 4-11.

Soil Survey for Scotland 1981. *1:50,000 Soil Maps of Orkney: Orkney Mainland, Orkney Northern Isles, Orkney, Hoy*. Macaulay Institute for Soil Research: Aberdeen

Vreeken W J. 1973. Soil variability in small loess watersheds: clay and organic carbon content. *Catena* 1: 181-96

Westeringh W van de. 1988. Man made soils in the Netherlands, especially in sandy areas (plaggen soils). In: van Waateringa W. Groenman, Robinson M (eds) *Man Made Soils.* British Archaeological Reports (International Series) 410, Oxford

Acknowledgements

Donald Davidson is grateful to the Carnegie Trust for the Universities of Scotland for a grant towards fieldwork costs in Shetland; Ian Simpson acknowledges Steve Dockrill for discussions on the Tofts Ness samples.

Chapter 5

Field Systems and Cultivating Implements

Alexander Fenton

In looking at the history of soil and field systems and their interactions with humanity, we are still finding a hesitant way though a subject area fraught with complexities that do not diminish but actually increase in number as we learn more. It is difficult to interpret the past on the basis of personally acquired or book-learned knowledge, and still feel sure that we are really understanding it. The further back we go, the more difficult it becomes. Nevertheless we have to try, even though all we may be doing is narrowing the parameters within which more informed interpretation may be attempted for given places and times. When we are trying to interpret medieval or later rural settlement by working backwards from the present or recent past, then we should also complement the task by working from the archaeological evidence of the remoter past forwards, taking into account whatever data can be gleaned from the evidence of early cultivation marks and ridges as they have turned up in places like Arran, in Perthshire (at North Mains and Strathallan), in the Borders, and elsewhere (cf. Barber 1982; Barclay 1983; Barclay 1989; Halliday 1986).

Of particular immediate interest are the plots of so-called cord rigs, each measuring about 1.3m broad, and ridges like those at North Mains which are about 0.15m high and 1.8-2m from crest to crest, since these give the possibility for a pincer approach from the past as well as from the present. For such an approach teamwork is essential. We do not as yet have in Scotland any such far-reaching and wide-ranging works as Mitchell's book on Valencia Island in Ireland (Mitchell 1989), and for lack of such a polymath we must continue to pool our knowledge.

What was happening, where, when and why?

If we think of the remoter past, cultivation was obviously more advanced at a much earlier period than the standard school text-books normally admit. It is, in fact, hard to get back to a stage of settlement where the alleged soil-scratching with a digging-stick was really found. Reasonably sophisticated forms of cultivation were being practised, seemingly throughout Scotland, very early on. Leaving aside the question of the marks of cultivating implements underlying the soil or in the sub-soil, and the ways that have been slowly evolved (to a great extent by Danish colleagues: cf. Nielsen 1970, 1993) to establish the direction of movement, the type of implement in use, and so on, I want to look first at the ridges themselves.

It seems safe to assume that ridges, with the associated furrows between, were a means of achieving surface drainage. They would not be expected, therefore, on well-drained, sandy, machair soils, for example, but more where the soils were damper, less permeable and more cohesive. Ridges and furrows can be made in two ways: with the spade and with the plough, though they are normally thought of as being made with the plough. As we know them, plough-induced ridges and furrows fall into two broad chronological groupings: those made by the old Scotch plough and those made by improved ploughs of the type developed by the Berwickshire plough manufacturer James Small, in the late eighteenth century. The type of implement used, and the size of the team, were intimately related to the form of the ridges and furrows:

> It was usual for writers of the Improvement period to speak of the old Scotch plough as a clumsy, awkward, inefficient instrument. It is true that the length of the sole combined with its flat mould-board created a lot of friction and made a large team necessary, but criticism should only be made within the framework of the conditions within which it worked, and of the cultivating techniques to which it was suited. In fact, the old Scotch plough was well adapted to the purpose it had to serve, which was to plough the soil into a series of ridges and furrows. The result was that the old Scotch plough always had to turn its furrow up the slope created by the side of the ridge, which in itself created further friction, and this also helps to explain the need for a large team. The plough, therefore, had always to be held tilted to the right, and in this position the sole acted as an anchor in the base of the furrow, whilst the flat mould-board, shaped to slope upwards from front to back, came into its correct position for pushing up and pressing into place the freshly cut furrow-slice. Such work could not have been easily done by light ploughs on the old high-backed ridges. . .
>
> The length of the team had in turn an effect on the configuration of the ridges. Ridges that are older than about 1750 are usually not straight, but have a serpentine appearance. This is almost certainly a result of two functional factors - firstly, the fact that the long team had to start turning on the headland before the plough was out of the furrow, and secondly the need to keep the mould-board pressed firmly against the furrow-slice right up to the end of the furrow. These factors could have had a cumulative effect over the years in giving the ridges a curved shape. There is, therefore, a close and inevitable link between the implement, its draught-power, and the old unenclosed plough-landscape, which had no dykes or hedges to inhibit the turning of the big teams. (Fenton 1976: 31, 35).

With Small's plough, and others developed from it to suit local soils, the high-backed, curving ridges of earlier days came to be straightened out, lowered in height, and worked in the form of parallel strips. The need for such ridges disappeared with the general adoption, from the second quarter of the nineteenth century, of systematic underground tile drainage (Fenton 1976: 18-23) that produced the pattern of level-surfaced, enclosed fields that has now been the standard farming landscape for over a century and a half.

The ridge and furrow systems we see surviving at the present time on the lower slopes of hills, on golf courses that preserve old farming landscapes and under plantations of trees, are the lowland equivalents of what are often now called lazy beds, made with the spade, in the more outlying zones. Which came first is a chicken and egg question, but one is an adaptation of the other, or rather, both are responses to specific land use requirements involving a need to have raised 'beds' with water drainage channels between. To ask how old lazy beds are is probably to ask how old ridge and furrow cultivation is. It is important to stress the quite sophisticated nature of lazy beds, as they survive or have survived till recently in places like the Faroe Islands (cf. av Skarði 1970: 67-73; Joensen 1980) and Ireland (Bell & Watson 1986: 43-63), as well in Scotland, involving team effort in their making, a variety of ways of handling the soil and the sods, and also a range of well-adapted spade and shovel types and clod breakers in the form of stubby-toothed wooden rakes or, in the Faroes, a flat wooden board or beater on a handle (Fenton 1994: 133-48) – which, by the way, no one would readily relate to a cultivating technique if they found it out of context.

This example, of ridges and furrows is an important one that had a vast spread whose extent has been greatly obscured by the impact of improved forms of agriculture and new plough types following the late eighteenth century. The resulting level or at least non-ridged fields to which we are now accustomed are nothing like the landscapes which our forebears knew.

Can zones of social/agricultural/economic practice be recognised?

In terms of ridge and furrow, zoning can be identified. They reinforce the Highland Line concept, at least in broad terms, being symbolised by the emphasis on the spade on the one side, and on the plough and its team on the other, though the Highland line is really far more like a sausage than a line, an elongated zone with a good deal of penetration of aspects of material culture – as of language and lore – from both sides.

Such Highland-Lowland zoning, however, leaves open the question of how widespread was the use of the spade in earlier times, when what happened in the Lowland areas may have been much more like what happened in the Highlands in more recent times. We need, through archaeology, to adopt a diatopic approach, to check if possible what was happening in different areas of specific bands of time in the past, as narrow as we can make them, before we can be sure that the zoning we think of naturally now was valid always.

What still survives and where?

Prominent survivals are the lazy beds of the margins of Atlantic Europe, and preserved traces of plough-induced ridges and furrows in the Lowland areas. The latter are often usefully portrayed in Slezer's illustrations of the late seventeenth century (Pl.5.1; Cavers 1993). Here there are indications of the blocking of ridge and furrow units into groups that may be seen as incipient 'fields', if we assume that a 'field' is a piece of ground worked by a single farmer, with or without a visible boundary round it. It must, however, be recognised that this concept of a field may not be the same as that of subsistence farmers in a joint-farming situation, where intermingled use of strips or patches was the norm, and the 'field' would have been the totality of these.

Pl. 5.1: A *bandwin* team of seven. Six people shear with the sickle on two rigs, and one man binds and stooks (Stephens)

What is the nature and potential of documentary sources for enhancing understanding of both general historical processes and extant field remains in particular?

It is obviously essential to seek out all possible sources of information, remembering always that documentary sources can be coloured by the message that the writer was trying to convey, or by failure to understand tools, techniques or processes that were not familiar. In many ways more essential is the effort to built up a thorough understanding of the functioning of rural communities, their resource needs and the pressures on them of those in authority, and also of the kind of skills through which their everyday existence was facilitated. The Danish experiment I want to mention later was a good test-bed for the last point.

Rather than try the impossible task of a full survey of the evidence (cf. Bangor-Jones 1993: 36-42), I shall focus on points that flow from documentary evidence. If we consider the forms of land use of the farming communities which were beginning to appear in details on the printed page in the seventeenth century, there can be no doubt that much change was incipient or beginning to take place, partly stimulated by growing population pressures. The spread of settlement was still proceeding; for example, Gordon of Straloch, writing about Strathbogie, noted how, before 1662, husbandmen had been gathered in farming villages, each with as much land as could be tilled by four ploughs. Continual woodland clearance, however, had so expanded the cultivable area that four ploughs were no longer enough. As a result, the old, clustered ferm-toun type of settlement was partly given up in favour of farmers settling on new pieces of land, occupied as separate units (Mitchell 1907: 267ff). We do not know how widely this same pattern of movement was occurring, but the example shows that change was in the air. We are not dealing with static concepts, and have to accept that there was a variety of local land use systems, which we must learn to identify and date.

However, all change was constrained by the nature of the resources and of the available technology. Regardless of land allocations within any one system, the standard features were first the strips and patches or rigs of arable adjacent to the houses, commonly called infield or 'muckit land', because it got all of what might be described as the 'made' manure; second, the somewhat remoter outfield that was primarily used for grazing and secondarily for cropping after animals had been folded on specific parts; and third, the rough grazing and moorland areas that also provided important resource materials for fuel, building and manuring (e.g. by means of composted turf middens). Oats alone were grown on the outfield. Since its organisation demanded the moving of stock from fold to fold followed by periods of cropping on the folded parts, the form of use of the outfield may be seen as a kind of shifting cultivation. It is a matter of intense interest and one for detailed further study that amongst the first of the improvement were those on the outfield. The infield was too precious to take risks with, but from the seventeenth century onwards outfield areas were gradually being brought up to infield standards through the application of lime, and played a strong pioneering role in the spread of permanent arable around the farms. This consideration could well be applied in the interpretation of estate plans, through which, for the later eighteenth and nineteenth centuries, a good visual picture of the landscape can be got, at least in places.

But meantime, though what I have been saying stems from a strictly functional view of the requirements of the older farming communities, I do not deny that we are not yet very clear about the origins of infield-outfield, nor about the distribution of the system, say as between the more pastoral and the more arable areas of the country. It is not certain that the distinction between infield and outfield was equally valid everywhere, and it would be wise not to see the subject in terms of too mutually exclusive options. But the overall point, using infield-outfield as an example, is that we need to use the documentary sources far more thoroughly, and with a close eye to regional variation, as a major means of establishing what existed in documented time, what its nature was, and whether it reflects continuity with the past or not. For this, the normal kinds of archival records, published sources, place-names and linguistic data, not to mention aerial photography, all need to be harnessed.

A recent Danish ploughing experiment

A major ploughing experiment was carried out over the last 15 years by Grith Lerche, Secretary of the International Secretariat for Research into the History of Agricultural Implements, and main editor of the journal *Tools and Tillage*. She accumulated information about all surviving parts of cultivating implements in Denmark and the surrounding countries, and all available archaeological evidence about the traces they had left in the soil or on the landscape. Using surviving plough parts as a basis, she reconstructed a medieval wheel-plough, and so disarmed Carlsberg Brewery that they provided horses and drivers to draw it for the years of the experiment (Pl. 5.2). Amongst her aims in using the full-scale reconstruction were:

1. To study and register the processes of wear on all parts and to test them against surviving parts discovered in peat bogs.

2. To test the differential rates of wear on the different parts of the plough. It turned out that though a beam might be in use for upwards of 100 years, other parts, especially those in contact with the soil, might be worn out or broken within days or weeks or within one or two ploughing seasons. Therefore a plough is a chronological composite, with a possible time span of up to a century.

3. (a) To establish an analytical basis for the interpretation of ridge and furrow field systems of long, narrow strips, and assess the extent to which they had been influenced and shaped by the ploughing implement and the actual ploughing method.

    (b) To see if it would be possible by ploughing alone to produce such ridged features, and assess the time required before any such features would be visible and measurable.

Pl. 5. 2: Experimental ploughing with a reconstructed medieval Danish wheeled plough, 1980 (Grith Lerche, Secretariat for Research on the History of Agricultural Implements, Copenhagen)

It is the question of ridge and furrows that interests us just now (though all parts of the experiment interconnect). The outcome was that in five seasons, a distance of 287km was ploughed, covering a field of 2¼ha that was divided into 15 strips, each 12m broad. The rate was 1½-3 strips a day, and a round of c.225-250m normally took 3-3½ minutes. Ploughing depth averaged 8-15cm, the average width of the furrows was 24-29.8cm, and the average number of furrows per strip was 41-42. Such technical date are important when we think of our own ridges and furrows, and, of course, one of the means of studying or own past landscapes is with the interpretative help of comparative data from elsewhere.

Grith Lerche's experiments showed that it took relatively few ploughings towards the middle of the strip to raise a ridge visible enough to be contoured on a map by surveying. As compared with fossil ridge and furrow, the experimental ridges were sharp and crisp; in fossil ridges, water percolation, root disturbance and soil creep have blurred the picture. Amongst the conclusions reached are that farmers tended to make shallower furrows before the period of improvements, confirming what is known from English sources (Rogers 1903: 75). This, if nothing else, conveys something of a different attitude to ploughing, though what it means in practical terms remains to be assessed. There is a tremendous amount of food for thought for us in this work (Lerche *in press*).

Conclusion

What I am advocating is an open-minded approach stemming from the functional needs of farming communities in different parts of Scotland, taking into account such factors as the technology available at different times, rent demands in kind, the spread of settlement due to population pressures, and finally, a willingness to accept that what we think we understand – eg the definition of a field – may not have been the same to our ancestors.

Bibliography

Bangor-Jones M. 1993. The Incorporation of Documentary Evidence and other Historical Sources into Preservation and Management Strategies. In: Hingley R (ed). *Medieval or Later Rural Settlement in Scotland*. Historic Scotland, Ancient Monuments Division, Occasional paper Number 1, Edinburgh

Barber J. 1982. Arran. *Current Archaeology* 83: 385-63.

Barclay G J. 1983. Sites of the third millennium BC to the first millennium AD at North Mains Strathallan, Perthshire. *Proc Soc Antiq Scot* 113: 122-281.

Barclay G J. 1989. The cultivation remains beneath the North Mains, Strathallan barrow. *Proc Soc Antiq Scot* 119: 59-61.

Bell J, Watson M. 1986. *Irish Farming 1750-1900*. John Donald, Edinburgh.

Cavers K. 1993. *A Vision of Scotland. The Nation Observed by John Slezer 1671 to 1717*. HMSO, National Library of Scotland.

Fenton A. 1976. *Scottish Country Life*. John Donald, Edinburgh.

Fenton A. 1994. Clod-Breakers and Rollers, with special Reference to Scotland. In: Pöttler B, Eberhart H, Katschnig-Fasch E (eds). *Innovation und Wandel. Festschrift für Oskar Moser zum 80. Geburtstag*. Graz.

Halliday S. 1986. Cord rig and early cultivation in the Borders. *Proc Soc Antiq Scot* 116: 584-85.

Joensen J P. 1980. *Färöisk folkkultur.* Liber Laromedel, Lund.

Lerche *in press. Ploughing Implements and Tillage Practices in Denmark, from the Viking Period to about 1800, Experimentally Substantiated.* Royal Danish Academy of Sciences and Letters' Commission for Research on the History of Agricultural Implements and Field Structures. (1994)

Mitchell Sir A (ed). 1907. *Geographical Collections Relating to Scotland made by Walter MacFarlane.* Vol II. Edinburgh.

Mitchell F. 1989. *Man & Environment in Valencia Island.* Royal Irish Academy, Dublin.

Nielsen V. 1970. Iron Age Plough Marks in Store Vildmose, North Jutland. *Tools & Tillage* I:3: 151-65.

Nielsen V. 1993. *Jernalderens Pløjning. Store Vildmose.* Vendsyssel Historiske Museum.

Rogers J E Thorold. 1903. *Six Centuries of Work and Wages. The History of English Labour.* Swan Sonnennschein and Co, London.

Skarði J av .1970. Faroese Cultivating and Peat Spades. In: Gailey A, Fenton A (eds). *The Spade in Northern and Atlantic Europe.* Belfast.

Chapter 6

Budgeting for Survival: Nutrient Flow and Traditional Highland Farming

Robert A Dodgshon

Traditional farming systems are a well-worked theme, but there has been surprisingly little debate over what they involved in terms of nutrient flow. I suspect this silence stems from the problem of specification. Because so much has to be guessed rather than calculated, there is a danger of producing very speculative analyses. In reply, I would argue that despite these problems of specification, just being aware of the questions involved can still contribute to our understanding of traditional farming systems.

The problems and possibilities are well shown if we try to reconstruct the nutrient flow that underpinned farming in the western Highlands and Islands before crofting and the clearances. Traditional farming systems everywhere were caught within a nutrient flow trap, but I want to argue that those to found in the western Highlands and Islands before crofting and the clearances were caught within a particular kind of nutrient flow trap, one that typified other mountain or marginal areas in north-west Europe.

Not surprisingly, the key to understanding this flow trap lies in appreciating how local environmental conditions affected the problem. In a stimulating paper written back in 1980, the historian Chorley argued that for traditional farming systems in northern Europe, the key nutrient was nitrogen. Whilst such systems re-cycled nutrients via straw and feed, and added to the cycle by transferring nutrients from grassland and meadow, they were crucially dependent on the nitrogen available from within soils via biological fixation (Chorley 1981, 80-1). Furthermore, even for a nutrient like phosphorus, Chorley suggests that the amount available within the soil through mineralisation was sufficient to meet the needs of traditional farming without the need for extra inputs (ibid, 88). Such conclusions are highly significant for the Highlands and Islands because factors like heavy rainfall, leaching, waterlogging, acidity – all common problems in the region – inhibited processes like biological fixation and mineralisation and, together, form the main reason why soils across the region have an 'inherent poor fertility' (Bibby et al 1982, 127). In other words, townships across the region would have had significantly less nitrogen from biological fixation and phosphorus from mineralisation than Chorley had in mind. In addition to such problems, townships in the region also faced the still more basic problem of having only limited amounts of land physically capable of being cultivated. Estate surveys compiled in the years on either side of 1800 suggest that less than ten percent was classed as arable (Dodgshon 1992, 174). Much of the land available to townships comprised ground that was high, exposed, steep, broken by rock outcrops, waterlogged or peaty. Though townships made great efforts to cultivate this sort of land, by far the greater proportion of it lay beyond the physical and climatic limits of cultivation.

These constraints on cultivation – the low levels of key nutrients provided through processes like biological fixation and mineralisation and the high percentage of land that was physically unsuited to cultivation – undoubtedly affected the way in which traditional communities set about maintaining nutrient flow on their arable. In the circumstances, the logical response for Highland communities was to counter the low nutrient status of their soils by recycling as much nutrient as possible via straw-feed and by adding as much extra to the cycle by transfering nutrients from their meadow and extensive pastures via animal dung.

These, of course, were strategies adopted by traditional farming communities everywhere (cf. Olsson 1991, 300-5). As solutions though, they had limitations. They did not allow arable or output to be expanded indefinitely. A relationship existed between the amount of arable that could be maintained and the amount of manure available, the latter, in turn, being constrained by the amount of straw feed, pasture and meadow available. How these constraints operated varied between different environments. In particular, we need to draw a distinction between how they affected communities in Lowland Britain and how they affected those in upland or marginal regions like the Highlands and Islands. In the former, far more of the total land available to communities could be cultivated. This meant that under pressure of population growth, it was possible to have an extension of arable to the point at which pasture reserves were reduced below those needed to support arable. The result was a reduction in the flow of nutrients from pasture to arable and a decline in yields. In a recent review of the problem, the agronomist R Shiel has suggested that for optimum yields, traditional communities needed to keep no more than between 15-20 percent of the land under arable (Shiel 1992, 71-2). As the proportion of arable increases, yields fall away significantly (see Fig. 6.1). The economic historian, M Postan, argued that such a fall took place in southern England over the late thirteenth and early fourteenth century. Rapidly increasing population was seen as precipitating a state of land pressure, with an over-expansion of arable leading to soil exhaustion as the flow of nutrients from grassland became depleted (Postan 1974, 23-5; 57-71). Detailed analyses by H S A Fox have confirmed that some communities in southern England were cropping as much as 50 percent of their land by the late thirteenth century (Fox 1984, 119-59). As Shiel's figures make clear, this would have greatly reduced the transfer of nutrients from pasture.

It could be argued that given the vast reserves of non-arable resource, traditional communities in the Highland and Islands were unlikely to have faced the spectre of a Postan-type crisis. After all, even around 1800, when population was close to a peak, arable did not exceed ten percent of the land available. Seeing the balance between arable and pasture in simple gross terms though, glosses the problem. If we look at how traditional communities manured their arable, we find the standard procedure – in lowland as well as upland areas – was to apply the manure produced during the winter months. This required the close management of stock over winter, with some being kept indoors over night and, in some cases, during the day so as to accumulate their manure.

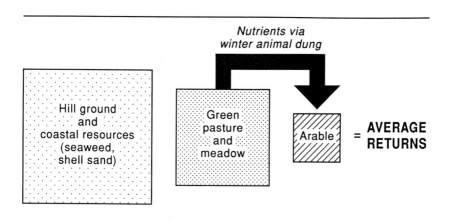

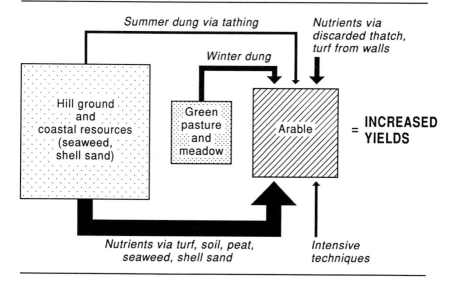

6.1: Effect of increases in arable on levels of nutrient flow from non-arable sources. In upper part of the diagram, balance between arable (15-20%) and green pasture/meadow (80-85%) enables substantial flows of nutrient via stock manure and hay so that yields on arable are at least average or above average. As arable is extended, levels of nutrient flow from green pasture/meadow fall away progressively. In the lower part of the diagram, though, areas with extensive acreages of non-cultivable land (i.e. hill or marginal ground) can offset this decline with strategies that make use of summer manure, the nutrients locked into resources like turf, peat, seaweed and shell sand, and more labour intensive techniques of preparing the soil. If exploited to the full, these strategies could lead to an increase in yields.

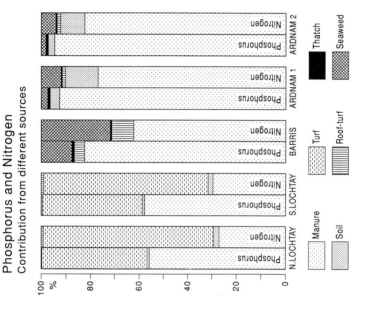

Fertilizers and Manure
Phosphorus and Nitrogen
Contribution from different sources

6.3: Inputs of nitrogen and phosphorus derived from all sources.

Based on Dodgshon & Olsson 1988: 47

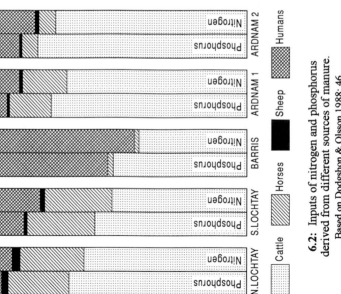

Manure Phosphorus and Nitrogen
Contribution from different sources

6.2: Inputs of nitrogen and phosphorus derived from different sources of manure.

Based on Dodgshon & Olsson 1988: 46.

Come spring, the manure so collected was then added to a part of arable.

This need to accumulate manure over winter has a critical bearing on the flow of nutrients available to communities in areas like the Highlands and Islands. As with all upland regions in north-west Europe, the number of stock which farming communities were able to keep over winter was determined not by the total amount of grazing available, but by their more restricted supplies of sheltered winter grazings, including the grass growing on balks between rigs, plus the winter feed supplement offered by hay meadows and straw. Some Highland townships were, in fact, divided into what they called 'winter ground' and outfield. These supplies of wintering ground and hay meadow had to come out of a very tight budget of 'better' land and to compete directly with arable for use of this 'better' land. When we add population growth and the consequent pressure for an expansion of arable, townships would have quickly come up against the limitations imposed by this limited budget of suitable land. Some surveys of Highland and Hebridean estates drawn up during the late eighteenth and early nineteenth centuries draw the distinction not just between arable and non-arable, but between what could be classed as green pasture and meadow and what was hill ground. If we take their supplies of green pasture and meadow as representing what was available to maintain stock over winter and compare it with arable, it soon becomes apparent that the arable-pasture ratios of Highland and Hebridean townships may have been far more adverse than what would be suggested by a simple reading of the overall balance between arable and non-arable (Fig. 6.1). In some cases, we are dealing with between 50-75 percent of 'better' land under arable and between 25-50 percent under wintering ground and meadow. Seen in these terms, the case for seeing traditional communities in the Highland and Islands as facing a Postan-type crisis by the late eighteenth and early nineteenth centuries becomes a little more persuasive. I should add that the poor straw value of Highland and Hebridean grains and the practice of graddaning, or separating out the grain by setting fire to the husk and straw, would not have helped nutrient recycling via straw feed. In short, if manure alone was used, population growth may have led Highland townships – like their counterparts in Lowland Britain – into a nutrient flow trap. Indeed, given the problems affecting the biological fixation of nitrogen and the mineralisation of phosphorus and the strong possibility that many Highland townships had less resources of winter pasture and feed, there is a case for arguing that potentially this nutrient flow trap was deeper than that facing their Lowland counterparts.

We can enlarge on the nature of this flow trap by looking at actual case-studies. A few years ago, I carried out a joint project with an ecologist, Dr. Gunilla Olsson, using data from four township clusters in the central and western Highlands (Dodgshon & Olsson 1988, 39-51). Using techniques developed by the Ystad project in southern Sweden (Olsson 1988, 123-37; Olsson 1991, 293-314), the study attempted to analyse the flows of nitrogen and phosphorus provided by animal and human waste. The study was based on calculations about the movement of stock between different sectors, on their stocking density, on weight/feed intake, the length of time manure was kept in storage, the loss of nutrients whilst in storage, and so on, in townships from four sample areas: North and South Lochtayside, Barrisdale and Ardnamurchan. As Figure 6.2 shows, different types of stock produced different quantities. Obviously, faced with this pattern of flow, a township cannot solve its problems by switching to stock that produce more manure for what matters is the constraint introduced by the total feed available. The souming system acknowledges this point. Souming was based on a weighted equalization of different stock (one soum = one horse = two cows = 5 sheep = 10 goats). Switching between stock does not alter the total amount of manure produced by the total number of soums available. What mattered was the total winter feed available and the total number of weighted or soumed stock this would support, or the number which a township could soum and roum in winter. One further point: as townships came up against the limits imposed by winterings and meadow, the simple fact of more people would have meant more nutrients, although the proportion of nutrients derived from human waste at Barrisdale involved very small acreages (Fig. 6.2). Systems with much larger acreages derived much less from this source. When we consider what the flows from all sources meant in absolute terms, they confirm that some townships might have faced a Postan-type crisis if they relied on the nutrients provided by animal manure alone, whether by recycling straw or by transfers from pasture and meadow. In actual fact, maintaining nutrient levels by conserving winter manure was only one of a number of strategies available to Highland and Hebridean communities. Though their nutrient flow trap was potentially a deeper trap than that of lowland townships, they had – like other upland or marginal areas in north-west Europe – escape routes. These escape routes depended on the simple fact that upland or marginal areas had large amounts of non-arable resource. This non-arable resource constituted a huge nutrient store, provided communities were prepared to explore new strategies for transferring it to arable. Three possibilities can be mentioned (see Fig. 6.1). First, there was outfield cropping. What distinguished outfield cropping was the simple fact that it depended on the use of summer manure, applied by a system of tathing. This meant that townships acquired a system based on the all-year-round use of manure. It directly addressed what contemporaries saw as the weakness of traditonal systems: the neglect of summer manure. Because it was summer manure, the quantities involved would have been good but not when set against the manurial cycle or frequency (one in 6-9 years) with which outfield systems were manured.

Second, it was possible for townships to exploit the vast resources of their non-arable sector in a more direct way. A range of possibilities existed: shell sand, seaweed, peat, peaty soil, turf, and heather via thatch. As inputs, these different transfers have varying significance for nutrient flow. Shell sand was used only in coastal townships and then only where calcium rich sand was available. Its main impact was in altering soil acidity, thereby increasing the availability of key nutrients. Seaweed was also a transfer used heavily in coastal and Hebridean areas (Fenton 1986, 48-82). As a source of nutrients, it tended to release nutrients quickly. Indeed, contemporaries were in no doubt that seaweed was essentially a quick-fix, one that could not sustain cropping for as long as animal manure. Significantly, it appears to have been associated with areas which practised a grass-arable rather than an infield-outfield system (Dodgshon

1993, 687-8). It did not do much for phosphorus levels but it has as much nitrogen as similar quantities of manure.

The organic transfers provided by the hill or waste ground of townships – peat, peaty soil and turf – form an interesting group. For inland townships, such as those on either side of Lochtayside, these may have provided a more important source of nutrients than animal manure. An indication of how much nitrogen and phosphorus these different types of nutrient transfer may have contributed is provided by Figure 6.3 which uses the same sample of townships as Figure 6.2. As can be seen, on the basis of the amounts assumed to have been used, turf may have accounted for as much as 70 percent of the inputs of nitrogen and just under 45 percent of the inputs of phosphorus. Clearly, these are significant amounts. Indeed, for Lochtayside, turf may have provided more significant transfers of these vital nutrients than livestock manure. A word of warning though, of all the figures that went into these calculations, those for the amount of turf transferred were the most difficult to derive in the analysis. It was used as a major source of nutrient transfer in various ways. First, it was a foundation for roof thatch. In this form, we find it re-cycled on a regular basis. Second, it was widely used for building temporary enclosures for outfield tathfolds. Third, it was as a manurial transfer in a direct way. Eighteenth-century sources for the southern Highlands suggests that the aggregate amount of turf cut for these various uses was considerable. What is more, from bylaws enacted, it is clear that some townships faced acute problems because tenants cut turf on good or green pasture not just on hill ground. An act of 1685 prohibited the cutting of turf in parts of the eastern Highlands (APS, viii, 494-5). Indeed, one local farmer argued that the turf spade had done more damage than the Act of Union! (Cameron 1873, 298). In the Hebrides, one or two estates (eg. Seaforth estate) issued regulations against the cutting of turf on good land but the most telling indication of the potential damage done by the cutting of turf and peat is provided the extent of skinned land. The damage done on Lewis was mapped for the West Highland Survey (Darling 1955, 272-8). As an input, the transfer of turf and peat clearly had an effect on soil nutrient status, but worked slowly. Turf especially, releases nutrients much more slowly than manure. Indeed, there is an ecological point in having turf locked into walls and dykes for some years, before composting it. Furthermore, its application in quantity leads to the gradual build up of organic matter or humus.

To judge from the figures available on yields, a third way in which nutrient flow could be favourably altered was by switching to more labour-intensive forms of cultivation. Using the spade and caschrom instead of the plough gave you a yield bonus. This was recognised by early commentators on the western Highlands and Islands. The spade gave an increase over the plough of about one-quarter and the caschrom an increase of one-third (Dodgshon 1992, 183). Where figures for returns on seed are available, they bear this out. Systematic data available for returns on seed in all the townships of Tiree during the mid-1760s, for instance, shows returns of 2.2 for oats and 3.5 for barley. By comparison, figures for Barrisdale – a far more marginal environment – suggest townships there had higher returns, about 5 (Dodgshon, 1993, 688-92). These differences can be explained partly by the fact that townships in the former mostly used ploughs whilst those in the latter – with only one exception – used the spade and caschrom. In fact, if the amount of nutrient inputted by manurial transfers and seed is set against that extracted as part of the harvested crop, it suggests that spade-based systems may have been the more efficient (Fig. 6.4).

My own view is that to see the use of hand tools as producing a more efficient mobilisation of nutrients is to see only part of the problem, since the spade and caschrom were labour intensive. So also was the transfer of manurial resources like peat, turf, seaweed and shell sand. I am inclined to see the problem in terms of a general strategy based on labour abundance, one that made use of both the spade and caschrom coupled with heavy transfers of manurial supplements. Together, they helped to counter the prospect of a Postan-type crisis in nutrient flow, turning the threat of a decline in yields per acre into one of increase. Of course, yields per acre were only part of the problem. There is a case for arguing that looked at from the point of view of subsistence as opposed to nutrient flow, Highland and Hebridean communities were in a no-win situation. Throwing labour at the problem on the scale required by the spade and caschrom, and by the extensive use of non-arable resources to supplement nutrient flow, probably led to a fall in yields per head so that when faced with a period of sustained population growth like the late eighteenth century, communities would probably have experienced greater not less pressure on subsistence. However, by raising yields per acre in a marginal way and by enabling cultivation to be extended out over difficult or marginal ground, including ground that could not be ploughed or cropped without considerable preparation, labour intensive techniques would have significantly dampened this pressure on subsistence. In effect, population pressure contained – to a degree – part of its own solution.

Once we see the problem in this way, then it opens up a further dimension. The sort of nutrient budget maintained by traditional farming would not have been a constant affair. Its character probably fluxed in step with population growth or decline and conditions of labour abundance or scarcity. When labour was scarce, we can expect standard manure-based systems, which, when coupled with plough-based agriculture, may have involved relatively modest levels of nutrient flow. During phases of labour abundance, though, we can expect two kinds of adjustment. A wider range of manures might be exploited and greater use might be made of hand-tools like the spade and caschrom. This would have meant that the flow of nutrients would have been increased and, arguably, the amount of nutrients extracted, or nutrient use efficiency, might have been increased, though probably at the expense of greater levels of environmental disturbance and, arguably, long-term sustainability. Clearly, this sort of fluxing runs counter to the Postan thesis. Instead of a nutrient flow trap that led communities into declining yields per acre, it supposes that communities in upland or marginal areas were able to adopt strategies that could lead to increased yields per acre in a Boserupian-like way (Boserup 1965). Indeed, it is now recognised that in parts of lowland Britain, some communities also worked to counter a Postan-type crisis over the late thirteenth and early fourteenth centuries by making use of new or more intensive strategies (Campbell 1983, 26–46).

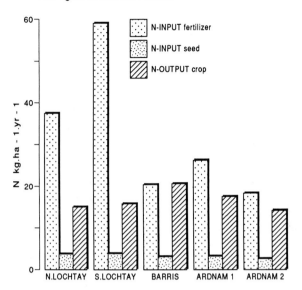

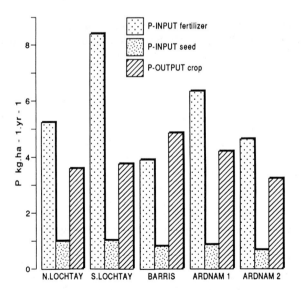

6.4: Nitrogen and phosphorus balance (based on Dodgshon & Olsson 1988: 48)

It does not follow, though, that all areas found their way out of this nutrient flow trap. Even in the Highland and Islands, it is possible to find areas where a case for declining yields per acre might be made. For instance, Duncan Forbes of Culloden's report on Tiree (1737) made out a strong case for soil exhaustion and reduced yields on the island (Forbes of Culloden 1737). Certainly, when we look at the figures available, the yields on Tiree were low. In fact, they are comparable with the figures which Titow produced for southern England during the late thirteenth and early fourteenth centuries, when pressure on land is also thought to have produced a decline in yields (Titow 1972). In the context of the Highlands and Islands, though, Tiree was exceptional. It faced a dilemma which no other Hebridean or Highland area faced. The proportion of land cropped by 1768 was very high, that is, 50 percent or 8240 acres. If Tiree did face problems of poor yields caused by an insufficient flow of nutrients, then it was perhaps simply because too much land was suitable for cropping. This might have led it into a much deeper nutrient flow trap than other areas. The Duke may have complained about the supernumeries on the island (Cregeen 1964, vol. 1: 1) but with so much arable, numbers may still not have been enough to intensify nutrient flows or techniques of cultivation in the way they were being intensified elsewhere in the Highlands by the mid-eighteenth century. But as the exception, Tiree may help prove the rule.

Bibliography

APS. 1814-74. *Acts of Parliament of Scotland*, 12 volumes. Edinburgh
Bibby J S, Hudson G, Henderson D J. 1982. *Soil and Land Capability for Agriculture: Western Scotland*. Soil Survey of Scotland, Macaulay Institute for Soil Research, Aberdeen
Boserup E. 1965 *The Conditions of Agricultural Growth. The Economics of Agrarian Change under Population Growth*. Allen and Unwin, London
Cameron A C. 1873. On ancient farming customs in Scotland. *Trans. Highland and Agricultural Soc. Scotland*, 4th series, vol. V: 296-9
Campbell B M S. 1983 Agricultural progress in medieval England: some evidence from eastern Norfolk. *Economic History Rev.*, 2nd series, XXXV: 26-46
Chorley G P H. 1981. The agricultural revolution in northern Europe, 1750-1880: nitrogen, legumes and crop productivity *Economic History Review*, 34: .71-93
Cregeen E (ed.) 1964. *Argyll Estate Instructions (Mull, Morvern and Tiree) 1771-1805*, 2 vols. Scottish History Society, 4th series, Edinburgh
Darling F Fraser (ed.) 1955. *West Highland Survey. An Essay in Human Ecology*. Oxford University Press, Oxford
Dodgshon R A. 1992. Farming practice in the western Highlands and Islands before crofting: a study in cultural inertia or opportunity costs? *Rural History* 3: 173-89
Dodgshon R A. 1993. Strategies of farming in the western Highlands and Islands prior to crofting and the clearances. *Economic History Review*, XLVI:.679-701
Dodgshon R A, Olsson E G. 1988. Productivity and nutrient use in eighteenth-century Scottish highland townships. *Geografiska Annaler*, 70B: 39-51
Fenton A. 1986. *The Shape of the Past 2: Essays in Scottish Ethnology* J. Donald, Edinburgh

Forbes of Culloden D. 1884. Letter concerning the Duke of Argyll's Estates in Tiree, Morvern and Mull, 24th Sept, 1737. Reprinted 1884 in: *Report of the Commission of Inquiry in the Condition of the Crofters and Cottars in the Highlands and Islands* XXX11-XXXV1, appendix A: 389-92

Fox H S A. 1984. Some ecological dimensions of medieval field systems. In: Biddick K (ed.). *Archaeological Approaches to Medieval Europe.* Studies in Medieval Culture, XV111. Medieval Institute, Western Michigan University, Kalamazoo: 119-58

Olsson E G. 1988. Nutrient use and productivity for different cropping systems in South Sweden during the 18th Century. In: Birks H H, Birks J J B, Kaland P E, Moe D (eds.). *The Cultural Landscape: Past, Present and Future.* Cambridge University Press: Cambridge

Olsson E G A. 1991. Agro-ecosystems from the Neolithic time to the present. In: Berglund B (ed.). *The Cultural Landscape during 6000 years in Southern Sweden - the Ystad Project.* Ecological Bulletins no. 41, Munksgaard, Copenhagen

Postan M. 1972. *The Medieval Economy and Society. An Economic History of Britain 1100-1500.* Weidenfeld and Nicolson, London

Shiel R S. 1991. Improving soil productivity in the pre-fertiliser era. In: Campbell B M S, Overton M (eds.). *Land, labour and Livestock. Historical Studies in European Agricultural Productivity.* Manchester University Press, Manchester

Titow J Z. 1972. *Winchester Yields. A Study in Medieval Agricultural Productivity.* Cambridge University Press, Cambridge

Chapter 7

Ridge and Furrow in the English Midlands

David Hall

Introduction

This chapter compares the remains surviving in the English Midlands, and examines threats to the survival of field systems and current action to ensure that significant examples are preserved.

The English Midlands form part of a zone running from north-east Yorkshire to Somerset, characterised by nucleated villages that were intensively cultivated before enclosure. In many cases up to 90 percent of the township area was arable, which necessitated careful regulation of arable and meadow, with communal pasturing of village animals on fallow land. The system was first fully described by Seebohm in 1883 and analysed nationally by Gray in 1915. The overview was brought up to date in 1973 with studies edited by Baker and Butlin.

The open and commonable fields of much of the English Midlands remained in use for most townships until the seventeenth century and for upwards half of them until the eighteenth century when they were enclosed by private Parliamentary acts.

In the champagne zone, the wide open expanse of arable was subdivided into many small, narrow arable strips called *lands*, that were grouped into blocks called *furlongs*. The furlongs were grouped into a few large areas called *fields*, which were open and hedgeless. The fields were generally cultivated on a two- or three-year rotation, one year being fallow. There were common grazing rights over the fallow at certain times. In the Middle Ages a farm, called a *yardland*, consisted of about 25 acres of land (the amount varied greatly from village to village), lying not in a block, but scattered in strips throughout the township, no two strips lying together.

Individual strips were probably first set out as a quarter or half an acre, but later, they often had average dimensions of about 7m by 180m, which is near to a third of an acre. The lands became ridged by the ploughing method, going around in a clockwise motion beginning at the middle, and finishing at the outside leaving a furrow. An anticlockwise motion was adopted in the fallow season to take some of the soil back and maintain a low ridge. The purpose of ridging was for soil drainage; the furrow acted as an open drain and as a clear demarcation between lands.

The ends of most lands are curved, the whole land taking the shape of a very elongated, mirror-image of an '*S*'. Curvature developed over the years, resulting from a tendency to draw to the left when performing a turning circle to the right. Maitland (1897, 379) drew attention to curved lands and they have been further discussed by Eyre (1955), who concluded that the ploughland shape was formed as a result of using a right-handed mould board.

The physical remains

The most readily visible effect of open-field strip ploughing is ridge and furrow (Pl. 7.1), but there are also other earthwork features, furlong boundaries being the most important. As well as moving soil towards the centre of the land, the plough moved small quantities in the direction of motion, towards the ends. This soil was left when the share was lifted out of the ground to turn. Over the years small heaps formed at each end, called *heads* first noted in records of the thirteenth century.

Two furlongs with lands lying in the same orientation had a boundary marked by a double row of heads, forming an irregular 'knuckle-like' bank called a *joint*. Where two furlongs had lands meeting at right angles the heads of all the lands in one furlong were piled on the first land of the next. These heads were ploughed over and smoothed out as part of the first land, which was called a *headland*. The profile and width of ridges vary, generally older ridge and furrow has a low profile.

Where a substantial amount of the pre-enclosure landscape is preserved, a variety of related features may be present. Roads and access routes to the fields can sometimes be seen as hollow ways, or there may be the mound of a windmill, located in an exposed position.

Headlands and joints survive in modern arable fields as banks of earth that represent the boundaries of the medieval fields (furlongs). Techniques of archaeological field survey used to reconstruct field systems, and the methods of furlong identification have been given (Hall 1982, 25-8). Much of the open-field landscape can be confirmed from the evidence of vertical RAF photographs taken in the late 1940s.

Records of open fields; field books

For many places there exists a wealth of documentary evidence that explains much of what can be seen on the ground. The most complete and important records of open fields are *field books* or parish terriers. These list every strip in a parish or township, grouped by fields and give details of individual furlongs and lands. Usually the quantity of land and name of the owner are stated; sometimes there is more information such as the type of tenure, the name of the tenant, and less frequently, the precise measurement of the lands. Some field books refer to maps and have summary tables listing tenants' names and various details about individual holdings. Field books afford a complete view of a township; field-system structure and furlong sizes can be established, along with details of land size, yardland size, land use, topography, tenure, and the spatial arrangement of the demesne and other estates.

Pl 7.1: Crick, Northamptonshire (NGR: SP 595 730), showing ridge and furrow which was in use until the area was enclosed by hedges in 1776

The major themes arising from historical studies of field systems are:

1   The identification of fields and furlongs.
2   Landuse; the disposition of arable, wood and meadow, and changes in the extent of arable. Late ploughing methods.
3   Information relating to the dating and origin of field systems.
4   Economy of field systems.
5   The manorial home farm, the demesne.
6   Details of regulations (bylaws) made to control communal village farming.
7   Farming methods.
8   The number of great fields and the overall cultivation regimes.
9   The work-service required by the lord to farm the demesne (before 1400).
10 The tenurial arrangement – i.e. how the lands were allotted, on the ground, to the lord of the manor and his tenants.
11 The structure and morphology of a field system.

Many of these items leave a physical record on the ground, particularly those arising from change of land-use (item 2), and are explained by the information from the open-field regulations and orders. Ground evidence also provides data for items 3, and 4.

Landuse and modifications in open-field practise since the high Middle Ages

There were several changes in the methods of working open fields from the fifteenth century until enclosure. Most of them were some form of contraction of the amount of arable and an increase in the grass area, changes made possible by the reduction of population that occurred during the fourteenth and fifteenth centuries, when less corn was needed.

The conversion of arable to pasture, for whatever reason, meant the cessation of ridging. Since the degree of ridging increased over the centuries, old enclosure or grassed-over lands are visible on the ground as low-profile ridge and furrow. The following types of arable contraction occurred.

Cow pasture
The most common type of increase of grass was to lay down arable by agreement to form 'cow pasture' for village herds. Examples of cow pasture occur in Northamptonshire from 1426 (Irchester), 1479 (Rushden), 1567 (Higham Ferrers) and 1712 (Ashby St Ledgers) (Hall *forthcoming*).

Leys
Parts of furlongs, and sometimes whole furlongs, were left to grass over permanently when they were called leys. Leys appear in the fifteenth and sixteenth centuries and continued in use until enclosure. The date at which they were mentioned in charters suggests a piecemeal formation from c1450-1550 (Hall 1982, Figure 20, shows the extent of leys at Ashby St Ledgers in 1715 and the

dates at which each block is first recorded). The extent of leys varies, being as low as three percent of the arable area in townships with good quality soils (Raunds 1739) but in less favoured townships there was up to 60 percent of grass (Arthingworth 1767; Hall *forthcoming*). Leys and cow pasture may be detected from the ground evidence as areas of ridge and furrow with a lower profile than that found in neighbouring furlongs.

Balks and grass ends
Narrow strips of grass, sometimes used as access routes, called *balks*, were formed by ploughing a few furrows away from the ridge of a particular land, leaving them to grass over. Balks are commonly recorded during the fifteenth century and later. Although reference to them occurs in field by-laws of 1300, examples dating from the fourteenth century are rare. They were used to mark out significant groups of lands, such as blocks of demesne lands, or major subdivisions of the fields.

By the seventeenth and eighteenth centuries deliberate attempts to increase the amount of grass in the open fields led many manorial courts to order that there should be balks or greensward between every land, the width of the balk to be roughly proportional to the width of the land. This type of balk was referred to as 'greensward' and would not have been common, but held with a particular land.

Grass ends.
Another late introduction of grass was by forshortening lengths of arable lands against a headland or joint. Instead of ploughing the full length of a land, several metres were left at the end, which would rapidly grass over. References to grass ends can be found in court orders dating from 1579 onwards (Hall *forthcoming*).

Late types of ridge and furrow
There are different types of ridge and furrow. All that so far described has been of medieval open-field type. There are also ridges produced after enclosure, when strips were ridged in the nineteenth century. These later ridges are very clearly distinguished from the older ones in that they are straight, parallel to at least one modern field hedge, and are usually wider (about 13m across) than the earlier, slightly curved lands. The furrows of this type of ploughing are picked up by a headland furrow that goes around the modern field. Straight-ridged lands can be seen at Kings Cliffe (TL 014 993), proved to be mid-nineteenth century in date because the ground was a wooded part of Rockingham Forest before then.

Another kind of nineteenth-century ploughing occurred leaving very narrow ridges only a few feet wide. Examples occur at Maidford in demesne closes that were once assarted pastures and never part of the open fields (at SP 60 52). Murray, writing for Warwickshire in 1813, notes that 'old ridges' vary in width but that new small ridges were two yards wide. Andrews (1853) and Stephens (1851) refer to then contemporary husbandry using narrow ridges. Examples of these two types of later agriculture also need preservation.

Dating and origin of ridge and furrow

There has been controversy in the past about the dating of ridge and furrow and whether it corresponded to that recorded on pre-enclosure plans. Map and aerial-photograph comparisons have been published by Beresford and St Joseph to show that in many places the comparison is exact (1979, 25-37).

Detailed comparisons for Northamptonshire show that in many cases ridge and furrow was in existence in the sixteenth century. Earlier dating can be demonstrated where there is coincidence of physical survival and particular events recorded in documents.

These cases illustrate the interaction of ground and historical evidence and the importance of surviving examples of ridge and furrow to improve understanding of medieval field- systems. Had the fields not survived then the evidence in the documents would have been of less value.

The origin of ridge and furrow seems to be towards the end of the Middle Saxon period, *c*700-800AD. At this time there was a replanning of the Midland landscape with settlements becoming concentrated in one place and strips laid out (Hall 1981; Hall 1982, 43-55).

Preservation of medieval landscapes

It is essential to preserve substantial quantities of ridge and furrow by an adequate means. There is currently no established mechanism for their preservation. Non-statutory methods, such as the Register of Historic Landscapes, does not ensure any protection for field systems, because the main threat to ridge and furrow, ploughing, cannot be controlled. It is therefore urgently necessary to provide statutory protection under the Ancient Monuments and Archaeological Areas Act 1979. This protection has ensured that examples of prehistoric and Roman field systems have been preserved, and should be extended to medieval fields.

Among the reasons for preserving good examples of ridge and furrow are the following:

a) Amenity value. Ridged fields form varied landscapes of interest to the general public.

b) Educational purposes. It is essential for students of the medieval period to see and appreciate what the agricultural system looked like. Its operation is then much easier to understand.

c) Examples of strip fields are needed for future research, when new questions will require new types of data. Present day recording of field systems followed by destruction, will not suffice, as evidence will be irretrievably removed.

d) Where ridge and furrow lies next to a monument it enhances the setting.

e) Examples of the subtle modifications, such as the formation of balks, forshortening of lands, early intake of ridges to form medieval enclosures, would be difficult to appreciate if they do not survive.

In Northamptonshire destruction rates currently run close to three percent per annum, and only a few parishes have as much as 40 percent of ridge and furrow surviving. In as little as 5-10 years most or all significant examples are likely to be obliterated. It is clear that urgent action is required to preserve significant and extensive samples for future study. It is insufficient to make a record and then allow destruction by cultivation; earthwork samples are needed for educational and amenity purposes and for future research. Criteria have been given for sample selection in Northamptonshire (Hall 1993, 28-9).

Initial preservation through Countryside Commission schemes have been successful in taking 517ha of Northamptonshire ridge and furrow during 1991-2. This forms a useful beginning, but is not a permanent solution, only a holding operation. It would take six years to include all the 3,000ha of the best preserved and documented field systems so far identified, but the scheme only operates for 10 years. More permanent protection is needed by means of the Ancient Monuments Act.

In the past field systems have not been scheduled in England because archaeological structures would not be found under them. Opinion is now moving towards regarding them as earthworks in their own right, worthy of preservation. They are an integral part of the economy and history of their related settlements and should not be isolated from them.

Currently, English Heritage has commissioned a new methodology to classify medieval vills with a view to defining regional parameters and select representative examples of all the types in England. It is proposed to link field systems to this work, likewise defining various types and selecting some of the best examples for preservation.

Bibliography

Andrews G H. 1853. *Modern Husbandry*. London
Baker A R H, Butlin R A. 1973. *Studies of Field Systems in the British Isles*. Cambridge University Press, Cambridge
Beresford M W, St Joseph J K S. 1979. *Medieval England*. Cambridge.
Gray H L. 1915. *English Field Systems*. Harvard.
Eyre S R. 1955. The curving ploughland strip. *Agricultural History Review*. 3: 80-94
Hall D N. 1981. In: Rowley R T. *The Origins of Open-Field Agriculture*. London: 28-32
Hall D N. 1982. *Medieval Fields*. Shire Publications, Aylesbury
Hall D N. 1993. *The Open Fields of Northamptonshire, the Case for the Preservation of Ridge and Furrow*. Northamptonshire County Council
Hall D N. *forthcoming*. *The Open Fields of Northamptonshire*. Northamptonshire Record Society
Maitland F W. 1897. *Domesday Book and Beyond*. Fontana, 1960
Murray A. 1813. *A General View of the Agriculture of the County of Warwick*. London.
Seebohm F. 1883. *The English Village Community*. London
Stephens H. 1851. *The Book of the Farm*. Edinburgh

Acknowledgement

I am grateful to Glenn Foard and the staff of Northamptonshire Heritage for providing information, and for their interest in preserving a significant part of the Midland landscape.

Chapter 8

'Gooding the Earth': Manuring Practices in Britain 1500-1800

Donald Woodward

Agricultural writers in early-modern Britain were well aware of the need to nourish the soil with a wide range of manures, but there were major problems when farmers attempted to put their advice into practice. Before the growing availability of commercial fertilizers in the nineteenth century – and they included both natural materials, such as South American guano and the bones of European battlefields, and the phosphates and nitrates produced by the developing chemical industries – many farmers were thrown back on the animal resources dropping all around them. Additionally, energetic agriculturalists could mix different soils together, bring in materials from outside – including seaweed, sand, chalk, and lime – or add the refuse generated by urban dwellers and industrialists. However, manuring was a matter of such routine that it has received inadequate attention from historians, rich though the subject is in colourful anecdote and juicy illustration. In part this is because it has left relatively little trace in farm documents, although some Scottish sources have recently revealed a rich harvest at the hands of Robert Dodgshon (1993).

For England more is known about attitudes to manuring among agricultural writers than among farmers, and a fuller rehearsal of these views has been made recently elsewhere (Woodward 1990). The two earliest authors, Fitzherbert and Tusser, were relatively silent on the issue, although both were well aware of its importance. Subsequent writers were less restrained and during the sixteenth and seventeenth centuries they mentioned almost every conceivable natural material as a source of nourishment to the soil. To give just three examples: in 1649 Walter Blith mentioned 'liming, marling, sanding, earthing, mudding, snail-codding, mucking, chalking, pigeons-dung, hens-dung, hogs-dung, or by any other means, as some by rags, some by coarse wool, by pitch marks, and tarry stuff, any oily stuff, and many things more, yea indeed anything almost that hath any liquidness, foulness, or good moisture in it, is very natural enrichment to almost any sort of land'; four years later Ralph Austen recommended that 'dead dogs, carrion, or the like, laid or put to the roots of trees . . . is found very profitable unto fruit bearing'; in 1677 John Worlidge suggested that dried tobacco was an excellent manure, and added 'which use is better than that to which it is now usually put unto' (Austen 1653; Thirsk 1983).

Dozens of treatises on agriculture and horticulture were published before the end of the seventeenth century and their authors invariably devoted some attention to manures, both ordinary and exotic. Authors borrowed extensively from each other, from the writings of their continental counterparts, and from classical writings, some of which were filtered through the continental literature.

However, many had a close acquaintanceship with agriculture and it has been argued that they were 'not hacks and plagiarists': indeed, 'their books combined the wisdom of their practical experience and their reading of others' (Thirsk 1983). Nevertheless, it is difficult to exaggerate the extent to which English writers were in thrall to their classical masters: as early as 1568 Thomas Hill listed twenty classical authorities in his *The proffitable Arte of Gardening* besides 'sundry others, whose names be here for brevity omitted'. Perhaps the best example comes from Robert Plot's *Natural History of Staffordshire* published in 1686. On his travels Plot saw a farmer ploughing in vetches which he regarded as 'the oddest sort of manure that ever I met with', but back at home in his library he soon found reassurance: 'upon consultation found it to be no new thing; Varro and Palladius both acquainting us of old, that they did not only plough in vetches to fertilize their land . . . but also lupins, and sometimes beans, for the purpose' (lupins are threatening to make a comeback as a field crop in the 1990s). The discovery that Greek and Roman authors waxed lyrical on the subject of manure did much to encourage English writers to do likewise, although some of their suggestions were of dubious value in an English context. Both Thomas Hill and Barnaby Googe (a sixteenth century author who brought Konrad Heresbach, the influential German writer, to the attention of English readers) followed classical authority in advising farmers to dung their land when the wind blew out of the west and the moon was decreasing (Woodward 1990): this advice was based on the rather dubious ancient belief that the waning moon helped to free crops of weeds, and that the wafting of the west wind in February was the harbinger of spring! A rather unhelpful suggestion for English conditions.

At its best – in the hands of influential authors like Gervase Markham, Walter Blith, and John Worlidge (Thirsk 1983) – the agricultural literature of the seventeenth century had a great deal to offer agriculturalists, although the impact on practising farmers is difficult to measure. What we do know is that a trio of well-known seventeenth-century farmers rarely strayed beyond the resources of their farms and adjacent land. Henry Best, who farmed in the East Riding of Yorkshire from the 1610s to the 1640s, referred only to the carriage of yard muck to the fields and the folding of sheep on the fallows (Woodward 1984). His Berkshire contemporary, Robert Loder, was a little more adventurous: in addition to yard muck and folding he dressed his fields with pigeon dung (which was a great favourite of Thomas Tusser), and added 'black ashes' which he bought in, and the malt dust partly produced by his own kiln (Fussell 1936). In the later seventeenth century, Nicholas Toke, who farmed in Kent, used dung, lime, and marl on his land (Lodge 1927). If, as seems highly likely, these three men were more market aware than many of their fellow agriculturalists, their experience would suggest that few farmers of the sixteenth and seventeenth centuries would have been able to benefit from the acquisition of off-farm materials, especially those of the more exotic kind. There is also the question of whether or not practising farmers had ready access to the growing body of agricultural literature: Henry Best, a gentleman farmer who owned over a thousand acres, owned a copy of Tusser, but he did not mention any of the more recent literature available to him. It must be surmised that few farmers read such works, although they probably helped to inform the opinions of the well-to-do whose attitudes would trickle down to the humbler sort at markets and fairs. Indeed, for the later eighteenth century it has been suggested that smaller farmers 'were more impressed by practical examples than by books' and that Arthur Young 'still placed his highest hopes of innovation on gentry, through whom the peasantry might be persuaded' (Thirsk 1983).

James Donaldson was one of the first Scottish writers into the fray. He seems to have been little influenced by the classical tradition, although he was well aware of writings south of the border. In his *Husbandry Anatomised, or an Enquiry into the Present Manner of Tilling and Manuring the Ground in Scotland* of 1697 he suggested that:

> Though some curious persons recommend horn and hoof, blood and guts of cattle, and shells of fishes, and salt-petre etc. as very strong and durable nourishment for the ground; yet seeing these things cannot be had by everyone, I shall speak of those which may be had everywhere, viz. dung of cattle, ashes, lime, marl, and sea-ware.

He added that some manures were better than others and that certain manures were more suited to particular types of land. A few years earlier John Reid had made a similar point in his *The Scots Gard'ner* of 1683, and stressed the fact that he was writing from a Scottish perspective (even in England the times for planting mentioned in modern gardening books must be taken with a pinch of salt in northern counties). But he was as guilty as many English authors in suggesting manures which were out of the reach of the average Scottish gardener, and thus attracted Donaldson's opprobrium.

In the early eighteenth century the tradition of listing a wide range of materials which could be used as manures was continued by that great manuring enthusiast, William Ellis of Little Gaddesden. In his *Husbandry, Abridged and Methodised,* he hardly missed a substance referred to in previous centuries and added some new suggestions of his own, such as old thatch and dogs' dung. But his whole approach was essentially practical: he reported on what he did himself, and on what he saw around him, or on his travels. He was aware of the work of at least some of his predecessors, but only once in his long discussion of manuring practices did he refer to the classical tradition, and then only obliquely and critically: he claimed to know a man who sowed his seed 'in the promiscuous, old, Virgilian way'. It is difficult to date with any degree of accuracy the waning influence of the classical tradition, but until the eighteenth century there were few who were prepared to flout ancient wisdom. Although he accepted that attention to wind and moon conditions was necessary when dunging, the German author Heresbach had recommended the application of dry dung, since 'though Columella do bid the contrary, our own experience wills us not to follow him' (Heresbach 1614). Similarly, John Evelyn refused to countenance the use of human excrement although 'mentioned by Columella', since it would taint any vegetation (Evelyn 1676). But, for the most part, Gooch, Evelyn and others were prepared to bend the knee in obeisance to the classical tradition. Proof that it was alive and well came in John Beale's *Herefordshire Orchards* of 1657 in which he expressed the hope that someone 'would publish a well-corrected copy of the four Roman husbandmen'. The crucial break from at least a theroretical dependency on the classical tradition occurred some time before Ellis began writing, probably in the later years of the seventeenth century.

For late eighteenth century and early nineteenth century England we can benefit enormously from the *General Views of Agriculture* for each county published between 1793 and 1817. They fall into two distinct groups – short pilot surveys ordered by the Board of Agriculture, most of which were published in 1794, and a more thorough second series, written to a standard format laid down by the Board and based in part on information submitted by readers of the first series. The complete absence of references to the classical tradition in the *General Views* marks them off sharply from the literature of previous centuries. They relied less on inherited lore and more on practical experience and observation (for full references to the *General Views* see Woodward 1990).

Both series of *General Views* show that certain manuring practices were common: yard muck is mentioned in most accounts; lime is referred to, usually favourably, in three-quarters of the 1794 reports and nearly all the later ones; marl, not available in all regions, appears in just over half of the reports although, as with liming, comments suggesting limited application appear quite frequently. As always a vast array of alternative manures was sniffed out by the diligent reporters. They enthused about the application of town muck, woollen rags, malt dust, peat ash, rabbit dung, salt, seaweed, soapers' ashes, bones, sticklebacks, and the like. But, as with the earlier literature, it is impossible to say what proportion of farmers, or even what proportion of parishes, adopted a particular practice, and the problem is exacerbated because some writers concentrated on what they regarded as best practice and frequently failed to make more general comments. No doubt there were many energetic and progressive farmers, like the Nottinghamshire gentleman who declared 'I raise heaven and earth to make manure' (Lowe 1794), but the *General Views* also contain enough comments to make it plain that many farmers had a great deal to learn about nurturing their soils. In Leicestershire it was said that few farm yards were 'well adapted to the economy of the dung-hill': most were 'paved with a dip for the drainage to run away' so that much of the goodness leached out of the dung (Monk 1794). In north Yorkshire the exasperated observer noted that 'no branch of rural economy is managed with less attention or judgement, than that of making and preserving the manure produced upon the farms' (Tuke 1794). Hill farmers in the Cheviots often had at their doors 'immense dunghills, the accumulation of unnumbered years, probably centuries': however, some – wiser than their surrounded neighbours – '*ingeniously contrived* to build their houses near a "*Burn side*" for the *convenience* of having it [the dunghill] *taken away by every flood*' (Bailey & Culley 1797: the italics are in the original). Similar comments were made in a number of other counties and, even in Hertfordshire, a county long noted for its dependence on London muck, the business was 'indifferently executed on very small farms' (Young 1804).

Three interesting debates emerge from the *General Views*: the issue of paring and burning; what might be called 'the pigeon question'; and the use of that which cannot be given its vernacular name – it must be referred to as 'night soil', *stercus humanum*, or 'the residue of human banquets'.

Paring and burning is the process of skimming off the top layer of earth and vegetation, heaping it up to dry, and then firing it. It was an old process, which had been mentioned by Donaldson with approbation, and was mentioned in over 40 percent of the 1794 reports. It received a mixed reception. It was very much approved of in Gloucestershire (Turner 1794), but disapproved of in six English counties, while reporters sat firmly on the fence in a further ten. Because the Board of Agriculture asked about the practice after 1794, paring and burning was mentioned in most of the second series of *General Views*. Most authors recognised that it had a valuable part to play in preparing certain sorts of land for cultivation: in particular, it was regarded as especially beneficial for land overgrown with coarse and intractable vegetation. As one gentleman in the Cambridgeshire Fens put it: 'burning beats the mucking cart' (Gooch 1811). Similarly, in Berkshire, paring and burning was an 'ancient practice', considered indispensable for 'all sour, tenacious soils, and where brakes, furze, and coarse grass cover the surface': moreover, it was given the approval of the royal bailiff and practised in Windsor Great Park (Mavor 1809). In Gloucestershire it was regarded as 'almost essential to the very existence of the agriculture of this county' (Rudge 1807). The debate serves to remind us that one of the greatest of the problems faced by pre-modern farmers was how to cope with weed infestation. In an age innocent of herbicides farmers often stuggled in vain against the plague of weeds that beset them: they coped – inadequately – by fallowing the land, thereby subjecting it to repeated ploughings, and by employing small armies of women and children in the spring and early summer to lop or uproot the worst offenders. After harvest, when the stock were turned out into the fields, they did not get the meagre pickings available from modern stubble, but a rich diet of variegated weeds which had grown amongst the corn.

The pigeon question aroused stronger passions. Pigeon dung was a favourite of the classical writers and had been praised in England since the time of Fitzherbert. According to Ellis it was 'certainly very much coveted by all husbandmen', and it was probably in more plentiful supply by the early eighteenth century with the renewed interest in pigeon-keeping after the Restoration (Thirsk 1984-5). In the first series of *General Views* pigeons made only a few fleeting appearances, although they were complimented on the quality of their muck, and, a few years later, Arthur Young continued a long tradition when he declared that, 'I have not the least doubt of the dung only of a well-stocked pigeon-house paying more than the necessary interest for building the house; an object greatly deserving a gentleman's or a rich farmer's attention' (Young 1799). But his support for the pigeon was out of sympathy with the views of many other agricultural writers: in the second series of *General Views* they commented on the excellent quality of the dung, but condemned the birds for the damage they did to crops. Much of the profit going to the owners of dovecotes was held to be a 'kind of public plunder, being derived from the corn-stacks and fields of the neighbouring farmers'. Indeed, 'no one who wishes to be upon good terms with his neighbours ought to keep them', and, since they were 'entirely granivorous, and withal extremely voracious', those 'insatiate vermin' were estimated to consume nearly five million bushels of grain a year, valued at nearly one and a half million pounds. To offset against the devastating plunder, some authors stressed the value of pigeons in picking up the seeds of weeds, but only in a few Midland counties was this felt to outweigh the damage to crops (Batchelor 1808; Priest 1810; Vancouver 1808; Parkinson 1808, 1811; Lowe 1798). In the nineteenth century pigeon dung became less essential with the import of South American bird muck.

In the 1960s one eminent English agricultural historian suggested that before the eighteenth century 'human waste was usually eschewed and buried deep and well out of the way' (Kerridge 1967). Such attitudes can be found in the early literature. In the sixteenth century Thomas Hill felt that 'that which men make, although it be thought most excellent, yet is it not so needful to be desired', unless it was for very poor land, and he added that its hotness made it 'greatly disliked' (Hill 1568). His distaste was mirrored by Reynolde Scot's remark that some men would not dung their corn land 'with so uncleanly a thing' (Scot 1574). Perhaps the strongest condemnation of the use of human excrement came from the pen of John Evelyn in the seventeenth century. He feared that plants 'contract the smell and relish of the ferments, applied to accelerate their growth', and so argued that 'we omit to enumerate amongst our soils, *stercus humanum*, which . . . does, unless exceedingly ventilated and aired, perniciously contaminate the odour of flowers, and is so evident in the vine, as nothing can reconcile it' (Evelyn 1676).

In the following century Ellis described the proper manner for composting night-soil, since 'want of knowing how to manage this hot dressing . . . has discouraged many from using it' (Ellis 1772). Nevertheless, prejudice meant that much of the valuable material continued to be wasted: indeed, although it was 'a highly fertilizing manure . . . the whole which is supplied by the city of Gloucester is thrown into the Severn' (Rudge 1807). Despite the hostile comments it is clear that night soil and town muck in general had been used by English cultivators since at least medieval times, and that such usage had been reinforced by the enthusiasm with which classical authors recommended 'the residue of human banquets as one of the best manures' (Woodward 1990). Even though opinion was flowing in its favour much human waste continued to be flushed into adjacent water-courses or dumped outside the growing towns: in London it was estimated that 'ninety-nine parts in a hundred of the soil of privies is carried, by the common sewers, into the Thames; which is a very great loss to agriculture' (Middleton 1798). The loss was indeed very great. Arthur Young estimated that, 'if the farmer manages his necessary-house in such a manner as to suffer nothing to run off from it, and frequently throws malt-dust, saw dust, fine mould, or sand, into it, he may from a moderate family, every year manure from one to two acres of land'. Thus, if nothing were allowed to seep away, the two million or so families living in England in 1801 should have been able to fertilize between two and four million acres of land a year without too much of a strain. But much was lost since many were guilty of 'letting their chamber-pots be emptied anywhere but where they ought' (Young 1799).

In the case of England it is very difficult to say much more about the practice of manuring by the farmers themselves, although some insights can be gained from the remarkable *General Views* for Cambridgeshire and Essex written by

Charles Vancouver and published in 1794 and 1795 respectively. However, the situation is much better for Scotland where the excellent work of Robert Dodgshon has shown us what can be done by a careful use of estate records. In the Western Highlands and Islands communities struggled to support growing numbers within traditional agricultural structures. Livestock were important for producing market goods, but the whole strategy of farming systems was not to maximize livestock production: indeed, 'livestock served the needs of arable and took second place to it in any competing claim on resources such as labour or sheltered land'. Large amounts of labour and horses were thrown at the problem of how to maintain fertility on the relatively small areas of arable. This was done by applying dung and moving huge amounts of shell sand, peat soil, and, above all, seaweed. Against the backdrop of rising population it was 'a struggle of demography against topography', and agricultural systems were 'organised so as to maximize subsistence, not marketable produce' (Dodgshon 1993).

In late eighteenth-century Essex and Cambridgeshire conditions were very different. No doubt farming strategies varied from place to place, and there were undoubtedly pockets of subsistence production, but many farmers were urged to increase their production by rising prices and the demands of their landlords. It might be expected that in this region above all others the advice of centuries would have been heeded with farmers acquiring a range of off-farm fertilizers to enrich their soils. In the early 1790s Charles Vancouver set out to find out the answer to this and a host of other detailed questions. Among the authors of the first series of *General Views* only he gave a detailed description of the soils and, to a lesser extent, of the agricultural practices of each parish for which he could get information. His approach to the two counties differed: he divided the 398 parishes of Essex into 14 districts; he did not group the 142 Cambridgeshire parishes but provided fuller information for many of them (Vancouver 1794, 1795).

Vancouver had relatively little to say about dunging and folding, although he recorded large flocks of sheep in many places: clearly these practices were too common to need detailed discussion. For manuring he divided Essex into two broad regions: a maritime zone running along the coast and penetrating inland along the rivers, and a much more extensive inland zone. In the inland areas 'farmers depend chiefly on that which arises from their own lands', which included 'correcting the natural defects of the several soils, by mixing the opposites of each other together'. The only places in the inland zone to receive substantial quantities of off-farm dressings lay in the south-west corner of the county, 'within the reach of the London muck'. By contrast the maritime zone benefited from the widespread application of 'foreign manures'. Along the river Stour farmers mixed fresh soil with 'London muck' available 'at the wharf' for 15s. a waggon load, and it was used along the Colne 'where the distance from the wharf, or landing place, does not absolutely forbid it'. Similar substances and chalk were used all along the coast, although they did not penetrate very far inland: carriage costs kept the area benefiting from such manures to a narrow ribbon.

Vancouver was able to obtain information for 105 Cambridgeshire parishes, and he discussed the manuring practices of 82 of them. As a land-locked county it was thrown back on its own resources rather more than Essex, although he gives the impression that the mixing of soils, claying, chalking and liming were little used. However, the practice of applying 'foreign manures' was rather more widespread than in inland Essex. Vancouver referred to such manures in 48 parishes, although for 18 of them he added the qualification 'occasional', and for 14 parishes he specifically said that they were not in use. 'Foreign composts' were used especially around Cambridge where the annual influx of undergraduates helped to boost the supply of 'town muck'. From the manuring point of view the county was divided into two zones, the Fens and other areas. Outside the Fens, yard muck, the sheepfold, and foreign composts were commonly applied, although, despite the great detail Vancouver supplied, it is impossible to say what proportion of farmers used particular additives. In the Fens, off-farm dressings were not needed. He argued that the large herds of cattle kept on fenland pastures 'accumulate such prodigious quantities of manure, as to preserve the arable land in good heart'.

Despite the inevitable biases which appear in all contemporary accounts of farming practices, Charles Vancouver's descriptions provide a uniquely detailed insight into farming methods in the two neighbouring counties. The chief impression left with the reader is that the majority of farmers remained heavily dependent on the internal resources of their own farms. However, bought-in dressings were not unimportant, especially in Cambridgeshire, which, outside the Fens, was more heavily arable than much of Essex. Bulky additives – such as the Kentish chalk which sold freely in the maritime zone of Essex – were usually applied only within a few miles of water: elsewhere excessive carriage costs prohibited their use.

The impressionistic evidence on which this chapter is based suggests that by the eighteenth century, and especially with improvements in the transport network, the desire and ability to tap reserves of off-farm fertilizers had increased and probably played its part in improving arable yields, which undoubtedly rose during the period, especially in the last century under consideration. Through the increasing use of 'foreign composts' a number of problems pressing on society could be relieved at a single blow: the collection of various natural products, industrial refuse, and urban waste produced a cleaner environment, raised agricultural yields, and provided employment. However, despite the many thousands of words of encouragement flowing from the pens of agricultural writers, most farmers probably had little access to off-farm dressings before the nineteenth century, and the revolution in manuring practices was probably not complete until the present century. Exotic manures were not available everywhere. Only those close to the shore were likely to apply sea-sand or seaweed to their holdings in significant quantities. Many dressings were available only in a few places: 'slam', a waste product from the Whitby alum works was only used in the immediate area, and the Nottinghamshire gentleman who raised heaven and earth to make manure sang the praises of sticklebacks, but added, 'alas! I have not been able to get any for these last ten years' (Tuke 1800; Lowe 1794).

Further progress in the delineation of changing sources of manure before 1800 will come only through further investigation of practice at the farm level, or through the investigation of estate records in the manner of Robert Dodgshon. A wide range of potential manures was available to the pre-modern farmer, although problems connected with their collection and distribution often precluded widespread adoption. Overland carriage soon rendered the movement of manure uneconomic, and no doubt, conservatism and inertia often exacerbated the problem. Before the nineteenth century it seems likely that most farmers – like the peasants in seventeenth-century Lincolnshire – 'knew only four ways of fertilizing their land, to leave it fallow, or to apply animal manure, vegetable waste, or other substances drawn from the soil such as clay and marl' (Thirsk 1957).

Bibliography

Austen R. 1653. *A Treatise of Fruit-Trees*. Oxford
Bailey J, Culley G. 1797. *General View of the Agriculture of the County of Northumberland*. Newcastle upon Tyne. (hereafter reference to the *General Views* will be as follows: *G V Northumberland*)
Batchelor T. 1808. *G V Bedfordshire*. London
Beale J. 1657. *Herefordshire Orchards, A Pattern for All England*. London.
Blith W. 1649. *The English Improver, or a New Survey of Husbandry*. London.
Dodgshon R A. 1993. Strategies of Farming in the Western Highlands and Islands of Scotland Prior to Crofting and the Clearances. *Economic History Review*, XLVI
Donaldson J. 1697. *Husbandry Anatomised, Or, An Enquiry into the Present Manner of Tilling and Manuring the Ground in Scotland*. Edinburgh
Ellis W. 1772 edn. *Husbandry, Abridged and Methodised*. London
Evelyn J. 1676. *A Philosophical Discourse of Earth*. London
Fussell G E. (ed). 1936. *Robert Loder's Farm Accounts, 1610-1620*. Camden Society, 3rd ser., LIII. London
Gooch W. 1811. *G V Cambridgeshire*. London.
Heresbach K. 1614. *The Whole Art and Trade of Husbandry, contained in Foure Bookes, enlarged by Barnaby Googe*. London
Hill T. 1568. *The proffitable Arte of Gardening, now the third tyme set fourth*. London
Kerridge E. 1967. *The Agricultural Revolution*. London
Lodge E C (ed). 1927. *The Account Book of a Kentish Estate, 1616-1704*. The British Academy, Records of Social and Economic History, VI. London.
Lowe R. 1794. *G V Nottinghamshire*. London.
Lowe R. 1798. *G V Nottinghamshire*. London.
Mavor W. 1809. *G V Berkshire*. London.
Middleton J. 1798. *G V Middlesex*. London.
Monk J. 1794. *G V Leicestershire*. London.
Parkinson R. 1808. *G V Rutland*. London.
Parkinson R. 1811. *G V Huntingdonshire*. London.
Plot R. 1686. *The Natural History of Stafford-Shire*. Oxford.
Priest St.J. 1810. *G V Buckinghamshire*. London.
Reid J. 1683. *The Scots Gard'ner*. Edinburgh.
Rudge T. 1807. *G V Gloucestershire*. London.
Scot R. 1574. *A Perfite Platforme of a Hoppe Garden*. London.
Thirsk J. 1957. *English Peasant Farming: The Agrarian History of Lincolnshire from Tudor to Recent Times*. London.

Thirsk J. 1983. Plough and Pen: Agricultural Writers in the Seventeenth Century. In: Ashton T H, Coss P R, Dyer C, Thirsk J (eds). *Social Relations and Ideas: Essays in Honour of R H Hilton*. Cambridge.

Thirsk J. 1984-5. *The Agrarian History of England and Wales, V, 1640-1750*. Cambridge.

Tuke J. 1794. *G V North Riding of Yorkshire*. London.

Tuke J. 1800. *G V North Riding of Yorkshire*. London.

Turner G. 1794. *G V Gloucestershire*. London.

Vancouver C. 1794. *G V Cambridgeshire*. London.

Vancouver C. 1795. *G V Essex*. London.

Vancouver C. 1808. *G V Devon*. London.

Woodward D (ed). 1984. *The Farming and Memorandum Books of Henry Best of Elmswell, 1642*. The British Academy, Records of Social and Economic History, n.s., VIII. London.

Woodward D. 1990. An Essay on Manures: Changing Attitudes to Fertilization in England, 1500-1800. In: Chartres J, Hey D (eds). *English Rural Society: Essays in Honour of Joan Thirsk*. Cambridge.

Young A. 1799. An Essay on Manures. *Annals of Agriculture*: 33.

Young A. 1804. *G V Hertfordshire*. London.

Chapter 9

Manuring and Fertilising the Lowlands 1650-1850

John Shaw

For many centuries Scottish farming featured a continuous chain of interdependency involving livestock and crops. Livestock was sustained through the produce of the land – and land was sustained through the produce of the livestock. The second half of this cycle – the preparation and use of farm manure – forms part, but by no means all of my subject.

Under the standard model of pre-Improvement farming, in the seventeenth and early eighteenth centuries, the infield was cultivated continuously and received most of the available manure (Fenton 1976). Bere (a four-rowed form of barley) was especially favoured with manuring – and with other fertilisers. Oat crops were less likely to be manured. In the south and east, wheat and pease might be included in rotations on well-manured land. Although the process was not fully understood at the time, farmers may have known that a pease crop had a beneficial effect on the land. We now know that this, and other leguminous crops, can fix atmospheric nitrogen in their root nodules. The term 'muckit land', used in some areas to denote infield, suggests the importance of manuring in distinguishing infield from outfield.

The outfield was farmed less intensively. It was raided for turf – for use in building, animal bedding or as manure for the infield. The occasional cropping of parts of the outfield was preceded by a period during which livestock was herded and folded, in temporary enclosures, on the land to be cultivated.

Parts of the outfield might also be subjected to paring and burning – a kind of slash and burn husbandry, in which the turf was cleared with a breast-plough (flaughter), dried and burnt, leaving ash as a fertiliser for the crop which followed.

The stimulus towards Agricultural Improvement owed much to the need to maintain and enhance the manuring of land in areas where the old system was in crisis. As cultivation expanded at the expense of grazings, fewer animals could be supported. The loss of livestock meant a loss of manure, without which the land could not be maintained in good heart. The answer, which emerged during the eighteenth century, was to incorporate sown grasses and root crops, such as turnips and swedes, into arable rotations. Muck was just as important to Improved as to pre-Improvement farming. With livestock amongst the arable, the land had to be enclosed. But that's another story.

Direct manuring by grazing animals is only part of this story. Dung from housed animals, fed on harvested crops, also played its part. It is a little known but crucial fact that some animals produce a better 'end product' than others. Sheep dung was the most highly prized, followed by that of fowl, horses and cattle (Whyte 1979, 69). Prior to 1850 few pigs were kept, except as an adjunct to dairying or distilling.

Reading any text on agricultural buildings might give the impression that sheep were never housed (e.g. Fenton & Walker 1981); but this has not always been the case. Before the native sheep were displaced by commercial blackface and Cheviot flocks, it was customary for most farms to keep a few sheep. John Naismith, writing in the 1790s of Hamilton parish, harks back to those times.

> These sheep were constantly attended by a boy or girl during the day, whom they followed to and from the pasture, and penned at night in a house called a Bught, which had slits in the walls to admit the air, and was shut with a hurdle door . . .; the floor was littered from time to time with dry straw, or turfs dried and piled up in summer. These little flocks were the peculiar care of their owners. (*OSA*, I: 184).

There is little doubt that sheep housing had been practised for many years prior to the 1790s – in 1536, for example, the Court Book of Carnwath records the illegal construction of a sheep house on the common (Dickinson 1937). One of the principal reasons for housing sheep was to accumulate their valuable manure: manure from sheep cots or bughts was considered 'the most powerful' available (Leslie 1811). A cart load was thought equal to two loads of byre or stable manure, and was especially good for bere crops (Henderson 1812). George Robertson reckoned that sheep manure from summer folding and winter housing produced 'an advantage little less than the profit made otherwise by them' (Robertson 1829).

The Society of Improvers favoured the use of sheep cots. Maxwell describes how such a house might be floored, with layers of sand, clay, lime and bedding, so as to secure both dung and urine. And he draws on a little folk wisdom to confirm the value of sheep urine:

> The Country Man, to express his Opinion of the Excellency of it, when he sees an extraordinary Tuft of Corn growing, says, *The black Ewe has pished there.* (Maxwell 1757)

Andrew Wight, on his travels in the 1770s and 80s found Improved sheep houses in Angus and East Lothian. The farm of Dryburgh, near Dundee, had been brought 'into great order' by the use of sheep manure (Wight 1778-84, I: 310), whilst near his home in Ormiston, he found a partly roofed sheep court, of a type which continued in use in the Lothians into the nineteenth century (Ibid, IV: 422-3: see Callander 1988).

Later writers who ridiculed farmers for having housed their sheep were clearly ignorant of its full benefits (Graham 1928).

Doocots, and the depredations of the birds which they housed, were a favourite target for Improving authors. In one typical publication, it was claimed that a farmer 'can hardly be subjected to a greater plague than the near neighbourhood of a large pigeon-house' (Somerville 1805). Yet even here there was a silver lining, in the high quality manure which they produced – the forerunner of the South American guano which was to prove so spectacularly successful in the mid-nineteenth century.

Horse manure from the stables was the next most valued fertiliser. Improved stables were provided with hard standing, which helped ensure that little of manurial value was lost. From the stable the mixture of dung and bedding found its way onto the midden.

The midden was at once the shame and the salvation of rural Scotland. Onto the midden went the muckings out of the sheep cot, stable and byre, used building materials such as turf and thatch (enriched with soot from the fire), human excrement, household waste and anything else which came to hand. Successive travellers and Improvers described in horrified terms the proximity to the house door of this putrefying heap and the foul liquor which seeped from it. Yet, once transported to the fields in creels on the backs of women and horses, on sleds or in carts, it underpinned the fertility of the land and the productivity of the crops on which life depended.

Despite its comparatively low value as a fertiliser, cattle manure was, throughout the period, the principal ingredient in the midden. The byre was a traditional source, but during the nineteenth century the production of cattle manure, from turnips, grass and straw, was elevated to an art and a science. Where cattle continued to be housed in byre stalls, in the dairying west and the stock-rearing north-east, elaborate systems were devised. Additional openings were made to facilitate mucking out. The midden was brought under control and glorified with the name 'dungstead', with a roof to keep the rain off and ventilated sides to let it breathe. Liquid manure was channelled to a central tank, from which it could be pumped into a barrel and sprayed upon the fields. James Kininmonth, who farmed at Invertiel, near Kirkcaldy, 'had been strongly impressed with the great loss sustained by farmers from allowing the drainage of stables, byres, dunghills &c., to waste off by any natural outlet'. In 1831 he built a small liquid manure tank and, in 1839, a much larger version, in stone, some 72 feet long, 6½ feet wide and 8-9 feet deep. The liquid was pumped from the tank into 160 gallon barrels, mounted on carts, and sprayed onto the land, principally as a top dressing for grass (Kininmonth 1847).

The most ambitious farmer in this respect was probably a Mr Brown of Libberton Mains, Carnwath, who in the 1860s installed a steam engine to pump liquid manure from large tanks with a view to irrigating six acres of his farm. The scheme was not a success (Tait 1885).

The alternative system of cattle housing, in courts, developed in the arable east. Here cattle were over-wintered and fattened on turnips, using surplus straw as bedding and as bulk for manure. In their crudest form they were simply farm yards, enclosed by walls and buildings but not segregated from them. Gradually they were separated off, some shelter provided in open-fronted sheds and feeding passages with serving hatches added in order to improve efficiency. Some even had their own railway, a good example of open courts served by railway still surviving at Thurston Home Farm, Innerwick, East Lothian. In spring the cattle were sold on and the yards, now a foot or two deep in rich manure, were mucked out.

A long-running debate over the relative merits of roofed and open courts ended, about mid-century in victory for the pro-roof faction. Lord Kinnaird was one of the exponents of roofed courts. In 1850 he built an entirely roofed steading (probably on his Rossie Priory estate) and carried out tests on the comparative value of manure from this and from open courts. The former produced substantially heavier crops, both of potatoes and wheat, leading him to conclude that, of all the benefits of covered courts, 'the most remarkable result of my experience is in the value of the manure' (Kinnaird 1853). Manure production and conservation was, therefore, a major determining factor in nineteenth-century steading design.

Farmyard manure was available to all, but farms in certain favoured locations might enjoy access to other fertilisers. Any farm close to a town or city had a decided advantage. Here was a ready source of household and human waste, organic industrial wastes such as those from tanning or soap-making and, it should be remembered, the manure from a considerable number of urban stables and cow houses – all there for the taking.

In 1778 one farmer was carting 250 cart loads of dung from nearby Dundee (Wight 1778-84, I: 310). From the 1790s there are numerous references to the agricultural use of urban waste – in Cardross (from Port Glasgow and Greenock) (*OSA*, XVII: 211), near Alloa (*OSA*, VIII: 611) and Hawick (*OSA*, VIII: 522), in a circle of parishes around Glasgow (*OSA*, 249, 340-41) (including Cambuslang, where it was the principal manure) and similarly around Edinburgh (*OSA*, I: 217; V: 320; XVIII: 365; XIX: 585). In the 1840s waste from Edinburgh and Leith was being transported by canal to Uphall (*NSA*, II: 88) and by canal or sea to Polmont (*NSA*, VIII: 197).

Nineteenth-century attempts to clean up towns and cities through legislation merely formalised the practice. Instead of waste being dumped in the street, it was collected from ashpits (which included the contents of ash privies) deposited in the street and eventually taken away by a contractor, usually a local farmer, who paid for the privilege (Ferguson 1948; 1958). This oddly-termed 'police manure' was a much sought after commodity, which was particularly useful as a top dressing on clay lands (Ferguson, 1958,175). In Kirkwall, and doubtless elsewhere, it was sold by auction in the 1880s, fetching prices which some bidders found excessively high (*Orcadian* 1886).

In rural areas earth closets provided a rich manure for horticultural use, but in urban areas it failed to gain popularity – not surprising, perhaps: it was calculated at the time that a ton of dry earth would be required per person each year (Robie 1884).

A more direct arrangement applied in sewage meadows (the origin of sewage farms). In 1770, or thereabouts, a water meadow was laid out near Kirriemuir, irrigated by a burn from the Loch of Kinnordy, which was said to receive 'much filth from the town of Kirriemuir and from plash mills. . .' By 1813 some 35 acres were being thus irrigated and were let to graziers for summer pasture (ibid). Later in the nineteenth century there were sewage meadows below Kilbarchan (ibid), Maybole (*THASS* 1867) and Carluke (Barr 1884). But by far the largest and most celebrated – or perhaps notorious – were the Craigentinny meadows near Edinburgh. Here sewage from the Old Town found its way via the Cowgate and the aptly named Foul Burn to extensive irrigated meadows. The practice seems to have originated in the eighteenth century but became increasingly contentious during the nineteenth. Elsewhere there were sewage meadows in Piershill, in Dalry, in salubrious Grange and, most scandalously, within the royal precincts of Holyrood Park (Ferguson 1948, 158-60; THASS 1878).

Drainage of land, rather than of towns, was a recurrent theme in agricultural Improvement, and a process which gave access to marl – a limey substance composed of the remains of fresh water moluscs. Described in 1774 as 'that valuable fund for improvement' (Wight 1778-84, I: 281), marl improved soil texture and had similar properties to lime. It had been dug from pits over many years, but the drainage of lochs in the eighteenth century gave access to new and abundant sources. In Angus, between 1730 and 1790, the Lochs of Balgavies, Restennet, Forfar and Kinnordy were successively drained or lowered and their marl exploited (Headrick 1813). Similar, localised deposits were being tapped elsewhere in the 1790s – in Galloway, the Merse, Perthshire and West Lothian (OSA passim: Dodgshon 1978). As an alternative to draining, the loch might be dredged; the Bruce Hunter papers in Forfar contain an illustration of a marl-dredging boat, with detailed instructions as to its construction and use (NRAS).

The use of lime as a fertiliser has a long history, probably dating back to the late sixteenth century and possibly earlier. Timothy Pont, writing in the 1600s, noted that the lowlands of Cunninghame had been 'much enriched by the industrious inhabitants lymeing . . . ther grounds' (Whyte 1979, 201). A report on certain parishes, in 1627, found that liming in East Lothian and Berwickshire had brought about as much as an eightfold increase in land valuations (*Reports* 1835). Lime was extensively used in improving sour, acid soils, though it was less effective without drainage. It proved particularly valuable in bringing marginal lands into cultivation.

Throughout the seventeenth and early eighteenth centuries, lime-burning was almost exclusively confined to limestone-bearing districts in the central lowlands with good access to coal. The exceptions to this rule tended to be coastal sites, such as Boddin in Angus (established 1696) (NSA, XI: 254) or Mey in Caithness (established 1741) (Donaldson 1938) to which coal could be transported by sea; but for some time after 1750 lime for the north and east of Scotland was brought by sea from the Firth of Forth or north-east England (Anderson 1794; Roger 1794). In an attempt to overcome the fuel problem, a limestone crushing mill was tried on the Robertson of Strowan estate in the 1760s (Shaw 1984). More generally peat came into use in Highland districts, though the lower heat achieved made for an inferior product (Marshall 1794).

Early kilns took the form of earthen, cone-shaped clamps – examples of these can be seen near Auchencorth Moss, a major area of seventeenth century reclamation. Eighteenth- and nineteenth-century kilns were stone-built and brick-lined, of substantial construction, sometimes with multiple flues, vents and draw-holes. Generally speaking they were built against steep banks, which saved money on construction and made easier the task of loading the alternate layers of coal and limestone (Skinner 1969). In limestone-bearing areas of the Highlands, such as mid-Argyll, Strathtay or the Angus glens, small, round peat-fired kilns were constructed.

Besides marl, there was another lime substitute in the form of shell sand. But another coastal resource, seaweed – known also as ware, wreck or wrack – was much more widely used as a fertiliser. The use of seaware is of undoubted antiquity. A charter of Inchgarvie, Fife, dated 1491, refers to grants of 'wrak' and 'ware' (*RGSS*). In 1500 the taking of 'wair of the sey' for the lands of Scryne, Angus, is described as being 'usit and wont in tymes bigane' (*ALCSCC*).

Seaweed has the same nitrogen levels as farmyard manure, twice the potassium but only one third of the phosphates. It remained effective for only one, or at most two seasons, but was highly esteemed as a fertiliser for bere (four-rowed barley). One commentator, writing of Angus in the 1680s, noted how it 'occasions a great increase of cornes where it is laid' (Ouchterlony 1844). In the early nineteenth century several writers stated that farm rents in East Lothian were higher where there was access to seaware (Kerr 1809).

Wrack or ware could be cut with a sickle or, more usually, collected from the sea shore after storms had detached it from rocks. Once washed ashore, the ware had to be retrieved before being washed away. At such times there was a real sense of urgency. In seventeenth century Banffshire there were complaints that ware gatherers had broken the Sabbath by convening on Sunday evenings (Cramond 1891; 1896). Much later, in nineteenth-century Berwickshire, a farmer recalled how the estate had had to ban covert night-time collecting in order to give everyone a fair chance to benefit from this bounty (Wilson 1902). In Ayrshire, from 1796 to 1887, a formally constituted 'Wreck Bretheren Society' met regularly to regulate access to and the collection of seaware (Hewat 1894; *see also* Noble 1975). Perhaps the most vivid account is one from early nineteenth century Moray:

> Whenever sea-weed is washed up on the shore, every other operation on the farm in its vicinity is suspended, and men, women and horses are day and night employed in one great exertion, to secure the whole quantity which has been deposited (Leslie 1811).

Access to the shore might be by a long-established customary route – such as the Ware Road which meanders through the otherwise strict geometry of the Tyninghame estate, to the sands of Belhaven Bay. In early nineteenth-century Berwickshire, some farms had roads made at great expense, descending the crags to the shore. Here, as in many other places, the ware was transported on horse-back in creels (Kerr 1809), though carts could be used where access was easier: there is a surprisingly early reference to the use of carts (as well as sleds) in the 1660s, at Crombie, Fife (*RPCS*: see also Fenton 1974).

There were differences of opinion as to how the seaware might best be applied – immediately or after composting with other materials such as earth or stable manure.

The manure of farm livestock was always, and still is, available – both problem and solution. To judge by the huge slurry tanks still being built (one of the few categories of building for which grant is still available) it has become more problem than solution. Imported and artificial fertilisers were already beginning to appear by 1850. Since then, the older means of enriching the soil – urban waste, marl, locally burnt lime and seaware – have all been superseded.

Bibliography

ALCS. 1496-1501. Acts of the Lords of Council and Session in Civil Causes Vol II: 350
Anderson J. 1794. *General View of the Agriculture of the County of Aberdeenshire.* Edinburgh
Barr J. 1884. Town Sewage. *Transactions of the Highland and Agricultural Society of Scotland* (THASS) 4th Series XVI: 179
Callander R. 1988. Sheep Houses in Midlothian. *Vernacular Building* XII: 3-13
Cramond W. 1891. *The Annals of Banff*, p.13
Cramond W. 1896. *The Charters of Aberdour* p.421
Dickinson W C. (ed) 1937. *The Court Book of Carnwath 1523-1542.* Scottish History Society, Edinburgh
Dodgshon R A. 1978. Land Improvement in Scottish Farming – Marl and Lime in Roxburghshire and Berwickshire in the Eighteenth Century. *Agricultural History Review* XXVI: 1-14
Donaldson J E. 1938. *Caithness in the Eighteenth Century.* Moray Press, Edinburgh
Fenton A. 1974. Seaweed Manure in Scotland. *In Memoriam Antonio Jorge Dias* III: 147-86
Fenton A. 1976. *Scottish Country Life.* John Donald, Edinburgh
Fenton A, Walker B. 1981. *The Rural Architecture of Scotland.* John Donald, Edinburgh.
Ferguson T. 1948. *The Dawn of Scottish Social Welfare*.Nelson, Edinburgh
Ferguson T. 1958. *Scottish Social Welfare 1864-1914.* Livingstone, Edinburgh
Graham H G. 1928. *The Social Life of Scotland in the Eighteenth Century.* Black, London
Headrick J. 1813. *General View of the Agriculture of Angus.* Edinburgh
Henderson J. 1812. *General View of the Agriculture of Caithness.* London.
Hewat K. 1894. *A Little Scottish World.*
Kerr R. 1809. *General View of the Agriculture of Berwick.* London.
Kininmonth J. 1847. On the Construction of Tanks. *Transactions of the Highland and Agricultural Society of Scotland.* 3rd Series II: 292-98
Lord Kinnaird. 1853. On Covered Farm-Steadings. *Journal of the Royal Agricultural Society of England* XIV: 336-43
Leslie W. 1811. *General View of the Agriculture of Nairn and Moray.* London.
Marshall W. 1794. *General View of the Agriculture of the Central Highlands.* London
Maxwell R. 1757. *The Practical Husbandman.* Edinburgh
NRAS. 1783. National Register of Archives (Scotland) 1398: Papers of Bruce Hunter, Forfar Box 2, Bundle 47/15
NSA. 1845. New Statistical Account of Scotland. 15 vols. Edinburgh
Noble R R. 1975. An End to Wrecking. The Decline in the Use of Seaweed as a Manure on Ayrshire Coastal Farms. *Folk Life* XIII: 80-83
The Orcadian. (1886). 23 October
Sinclair J. (ed). (1791-99). OSA – *The Statistical Account of Scotland*, 21 vols, Edinburgh
Ouchterlony J. 1844. Account of the Shire of Forfar circa 1682. *Spottiswoode Miscellany* I: 311-50
RGSS. (1424-1513) *Register of the Great Seal of Scotland* Vol II, 429 No.2038
RPCS. *Register of the Privy Council of Scotland* 3rd Series I: 386
Reports. 1835. *Reports on Certain Parishes in Scotland 1627.* Maitland Club, Edinburgh
Robertson G. 1829. *Rural Recollections.* Irvine

Robie D. 1884. Town Sewage and its Application to Agriculture. *THASS* 4th Series XVI: 125

Roger, Mr. 1794. *General View of the Agriculture of the County of Angus or Forfar.* Edinburgh

Shaw J P. 1984. *Water Power in Scotland 1550-1870.* John Donald, Edinburgh

Skinner B C. 1969. *The Lime Industry in the Lothians.* University of Edinburgh, Edinburgh

Somerville R. 1805. *General View of the Agriculture of East Lothian.* London

THASS. 1867. Anon. *Transactions of the Highland and Agricultural Society of Scotland,* 4th Series I: 43

THASS. 1878. Anon. The Agriculture of Midlothian. *THASS* 4th Series IX: 24-5

Tait J. 1885. The Agriculture of Lanarkshire. *THASS* 4th Series XVII:18

Whyte I. 1979. *Agriculture and Society in Seventeenth Century Scotland.* John Donald, Edinburgh

Wight A. 1778-84. *The Present State of Husbandry in Scotland.* 4 vols. Edinburgh

Wilson J. 1902. Half a Century as a Border Farmer. *THASS* 5th Series XIV:40

Chapter 10

Long-term Consequences of Using Artificial and Organic Fertilisers: the Rothamsted Experiments

John A Catt

Abstract

The long-term field experiments on heavy loam soil at Rothamsted, Hertfordshire, on the nutrition of arable farm crops provide much evidence for the effects of artificial fertilisers and organic manures on soil properties, crop yields and plant assemblages. The experiments on continuous growth of winter wheat (started 1843) and spring barley (started 1852) continue today and show that with good husbandry yields can be sustained with either artificial fertilisers or manures. On plots receiving 35 tonnes (t) farmyard manure/ha/yr the soil organic matter continues to increase; on those receiving inorganic fertilisers or no applied nutrients (nil plots) soil organic matter contents have remained stable for many decades. In similar experiments started in 1876 on sandy soil at Woburn Experimental Farm, Bedfordshire, yields and soil organic matter both declined over 50 years; however, loss of yield could probably have been avoided by better husbandry. Soil acidification through use of ammonium sulphate as a source of nitrogen was more rapid at Woburn than at Rothamsted, and probably led to greater pest and disease damage. Soil acidification has also resulted from atmospheric pollution, and has been especially rapid under experimental woodland (Wilderness Experiments) because of the increased scavenging by trees.

Use of sewage sludge at Woburn Experimental Farm led to persistent toxic effects on plants and soil micro-organisms (especially symbiotic nitrogen-fixing bacteria) because of heavy metals (mainly zinc, cadmium and copper) in the sludges used. Analysis of archived soil samples from the experiments has shown that there has also been increasing deposition of heavy metals and various organic toxins from the atmosphere at Rothamsted over the past century. However, retrospective analysis of crop samples taken over this period shows that very little of these enter the crops.

Introduction

Artificial fertilisers, introduced in the nineteenth century and now used in large quantities in many parts of the world, have revolutionised agricultural production. Together with developments in plant breeding and protection of crops from pests and diseases, they offer at least the potential for adequate nutrition of the world's expanding population. However, there is widespread concern that long-term usage of artificial fertilisers and other modern farming techniques are irreversibly degrading the world's soil resources, rendering them less capable of sustaining the increased levels of production. Soil degradation can take several forms. The most obvious is wholesale loss of the soil material by accelerated erosion, a very real and common problem in many regions where arable farming has spread onto areas that are susceptible to erosion because of steep slopes, intense rainstorms, strong winds, soil freeze-thaw cycles or inadequate crop cover in critical seasons. Less obvious forms of soil degradation, such as weakening of the physical bonds holding soil aggregates together or chemical and microbiological changes that impair the complex processes of nutrient cycling and supply, are on the one hand more difficult to assess and on the other more likely to concern a partially-informed public; they just may be 'time-bombs' waiting to wreak havoc on our food supplies or on the environment at some time in the future. Evidence for or against such threats can be provided only by careful long-term field experiments, in which soil treatments with, say, artificial fertilisers are continued on a regular basis for many years and the effects on soil and crops measured frequently. The agricultural field experiments at Rothamsted Experimental Station, Hertfordshire, provide just this type of evidence, though this was not their original purpose.

Between 1843 and 1856 Lawes and Gilbert started long-term experiments at Rothamsted on nutrition of most of the farm crops grown at the time (Jenkinson 1991). A few were discontinued in the nineteenth century when yields became very small, probably because the crops suffered from pests and diseases that were uncontrolled in monoculture. However, six still survive in scarcely modified form because their yields have been maintained or improved. These are the Broadbalk Winter Wheat, Hoosfield Spring Barley, Alternate Winter Wheat and Fallow, Exhaustion Land, Park Grass Hay and Garden Clover Experiments. Together with two setaside experiments started in the 1880s (Broadbalk and Geescroft Wildernesses), they constitute the Rothamsted Classical Experiments. The Broadbalk Experiment has been suggested as the longest continuous scientific experiment of any type (Garner 1959; Catt & Henderson 1993). One of the many far-sighted decisions taken by Lawes and Gilbert with respect to the Classical Experiments was to archive crop and soil samples for future study; these have recently been used for purposes that Lawes and Gilbert could never have envisaged.

The changing crop yields and soil characteristics of the classicals and some later experiments at Rothamsted provide reliable evidence for the long-term consequences of repeated use of artificial fertilisers and organic manures, usually farmyard manure (FYM), on fine loamy to silty soils in lowland Britain. However, the results should not be extended uncritically to other environments. For example, interesting contrasts are provided by similar experiments started since 1876 on the coarse loamy to sandy soils at Woburn Experimental Farm, only 20 miles west of Rothamsted.

Cereal yields and soil organic matter content

On the Broadbalk Winter Wheat Experiment yields of the variously fertilised plots changed little over the period 1852-1968 (Fig. 10.1), though all declined for a while in World War I and the early 1920s because of difficulties in controlling weeds (Garner & Dyke 1969; Dyke et al. 1983; Johnston 1989). The previous hand-hoeing of the plots then became impracticable and in 1925 a bare-fallowing system was introduced to control weeds; since 1957 herbicides have been used. Following the introduction of short-strawed varieties in 1968, yields increased sharply, especially on plots receiving higher nitrogen (N) rates. In 140 years yields have not declined on the unfertilised (nil) plot, and there has been little difference between yields on the FYM plot and the best achieved with artificial fertilisers (for a long while 96kg N, 35kg phosphorus (P), 90kg potassium (K) and 35kg magnesium (Mg) per hectare per year). On this soil, wheat yields can therefore be maintained by artificial fertilisers without the need for organic manures.

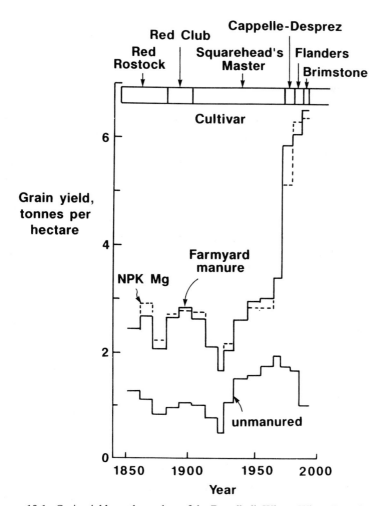

10.1: Grain yields on three plots of the Broadbalk Winter Wheat Experiment, 1852-1992

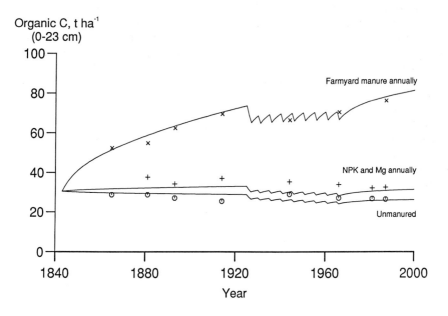

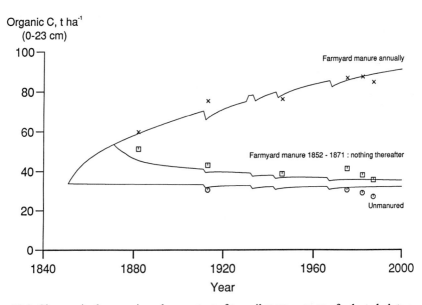

10.2: Changes in the organic carbon content of topsoil (0-23cm depth) of selected plots on the Broadbalk Winter Wheat Experiment (above) and the Hoosfield Spring Barley Experiment (below). The points marked are measured values, and the lines are calculated contents based on a model for turnover of soil organic matter (Jenkinson et al 1987)

After an initial small decline, the soil organic matter contents of the nil and inorganically fertilised plots of Broadbalk have remained constant (Fig. 10.2), the inputs from crop residues (mainly stubble and roots) approximately balancing losses by mineralisation and crop offtake (Jenkinson 1977). In contrast the organic matter content of the FYM plot has increased progressively over the past 150 years and is currently about 2.5 times those of the other plots. Since 1852 the patterns of yield differences (Warren & Johnston 1967; Johnston & Mattingly 1976) and changes in soil organic matter (Jenkinson et al. 1987) (Fig. 10.2) on plots of the Hoosfield Spring Barley Experiment have been broadly similar to those on Broadbalk. The plot receiving 35t FYM/ha annually now has over 3% organic carbon (C) compared with only 1% on the nil plot. This suggests that the maintenance of large cereal yields for long periods at Rothamsted does not depend on soil organic matter content.

The same originally did not seem to be as true at Woburn, where yields of wheat and barley grown continuously on the sandy soil declined progressively between 1877 and 1926 on nil, inorganically fertilised and FYM plots (Table 10.1), and over the same period there was also a rapid decline in soil organic matter (Johnston 1986; 1991). However, other factors may have been involved at Woburn. Loss of yield was largest on plots receiving N as ammonium sulphate, which has an acidifying effect especially on clay-poor soils. Tests of lime were started in 1898 (Johnston 1975) but did not fully restore the original yield levels. As there was less yield decline in other experiments at Woburn, where cereals were grown in rotation with other crops, the decrease may also have resulted partly from greater incidence of diseases or pests such as cereal cyst nematodes (Powlson & Johnston 1994), which proliferate easily in monoculture on sandy soils and persist even after liming has raised the pH.

One of the Hoosfield plots received FYM from 1852 to 1871 only, then no fertiliser until 1968 (Fig. 10.2). Between 1872 and 1968 the soil organic C content of this plot decreased from 2% to 1.5% (Jenkinson 1991), but the final value was still greater than that of the plot receiving inorganic fertilisers (49kg N, 35kg P and 90kg K per hectare per year) since 1852 (1.1% organic C). This indicates that some FYM components are very persistent in the Rothamsted soil.

Changes in soil organic matter in Ley-Arable experiments

Further evidence for rates of loss and accumulation of soil organic matter is provided by the Woburn and Rothamsted Ley-Arable Experiments started in 1938 and 1948, respectively (Johnston 1973; 1986). At Rothamsted, soil previously under old grassland and containing 3% organic C continued to accumulate organic matter slowly when kept in grass, but since about 1965 its organic C content has stabilised at 3.5%. When ploughed and kept in arable the same soil lost about 30% of its organic C between 1948 and 1980. Other plots at Rothamsted were established on an old arable soil with 1.7% organic C. Those subsequently kept in arable changed little in organic content, but that of a plot sown to grass and kept unploughed had increased by 30% by 1980. Yet other plots had 3-year grass or lucerne leys followed by test crops of wheat, potatoes and barley to measure the effect of leys on crop yields. Compared with continuous arable cropping, the leys did not significantly improve yields. Boyd et al. (1962) attributed this to the robust physical structure of the Rothamsted fine loamy soil, which provides a good rooting medium for arable crops without the need for improvements in structure and organic matter content resulting from the leys.

Table 10.1. Effects of N applied as ammonium sulphate or in farmyard manure and of liming on yields of winter wheat and spring barley (t/ha, grain) grown continuously at Woburn Experimental Farm 1877-1926 (Johnston 1975).

Crops and treatments	1877-86	1887-96	1897-1906	1907-16	1917-26
Winter wheat					
Nil	1.08	0.83	0.61	0.66	0.46
Farmyard manure[‡]	1.76	1.83	1.69	1.38	1.20
Ammonium sulphate[*], no lime	2.04	1.94	1.68	1.11	0.64
Ammonium sulphate, limed[†]	-	-	-	1.25	0.66
Spring barley					
Nil	1.56	0.98	0.60	0.60	0.49
Farmyard manure[‡]	2.39	2.30	1.87	1.87	1.54
Ammonium sulphate[*], no lime	2.57	2.10	0.19	0.19	0.30
Ammonium sulphate, limed[@]	-	-	1.39	1.39	0.90

[*]   46 kg N/ha plus phosphorus and potassium
[‡]   Mean of 17.6 t/ha/yr
[†]   2.5t CaO/ha, half in 1905, half in 1918
[@]   10.0 t CaO/ha, half in 1898, half in 1912

The Woburn Ley-Arable Experiment provided comparisons between continuous arable with inorganic fertilisers, continuous arable with FYM applied every 5 years and plots with a 3-year grass ley, 2-year arable rotation, also with and without FYM every 5th year. The most rapid increase in soil organic matter was on the ley-arable plot with FYM; those in continuous arable showed little or no change in soil organic matter over 33 years.

Taken together the results from Broadbalk, Hoosfield, the Ley-Arable Experiments and the various long-term cereal experiments at Woburn suggest that, with optimum inorganic fertiliser applications, liming to prevent acidity developing and a suitable crop rotation to avoid the worst effects of pests and diseases, soil organic matter and yields can be maintained on sandy as well as heavier soils. However, the standard of husbandry required is greater on sandy soils, where problems can develop more quickly and easily. Organic matter can be increased on both soil types by large (35 t/ha) applications of FYM every 1-5 years or by 3-year or longer grass leys, but returning grassland to arable results in a rapid loss of organic matter, especially on sandy soil. Under a continuing arable regime the soil organic content eventually stabilises, though at a lower level with inorganic fertilisers than with organic manures.

Causes and effects of soil acidity

Soils in the UK are progressively acidified by various processes, including decomposition of organic components, atmospheric deposition of sulphur and nitrogen pollutants, use of ammonium sulphate as a fertiliser and offtake of calcium by crops. In soils containing free carbonates or large amounts of clay there is considerable buffering capacity and the decrease in pH is slow, but soils with little clay and no free carbonate acidify rapidly. On the sandy, carbonate-free soil (Cottenham and Lowlands series) at Woburn, acidity accelerated by repeated use of ammonium sulphate caused recognisable problems within about 20 years (Table 10.1). However, the greater buffering capacity of the Rothamsted soils (mainly Batcombe and Hook series), with at least 25 percent clay and original pH of about 8.0 because of earlier chalking, delayed the onset of acidity problems here for at least a century. Through experience gained at Woburn the worst effects on cereal yields at Rothamsted were avoided by applying lime.

However, in the Four-course Rotation Experiment started in 1848 by Lawes and Gilbert on Agdell Field, Rothamsted, increasing soil acidity resulting from use of ammonium sulphate decreased yields of turnips after about 80 years (Table 10.2), and the experiment was consequently terminated in 1951 (Johnston & Penny 1972). Yields of other crops in the rotation (wheat, barley and legumes) were not so strongly affected (Table 10.2), and liming did not improve the turnip yields. This was probably because the turnips suffered from club root disease, which flourishes in acid soils and, once established, persists for long periods.

Table 10.2. Yields (t/ha/yr) of turnips and winter wheat in the Agdell Four-Course Rotation Experiment 1848-1951 (from Johnston & Penny 1972).

	CROPS AND TREATMENTS			
	Turnip roots		**Wheat grain**	
	Nil	NPK	Nil	NPK
1848 - 51	1.31	2.92	1.91	1.93
1852 - 83	0.24	3.47	1.46	1.96
1884 - 99	0.13	5.09	1.60	2.47
1900 - 19	0.11	3.98	0.97	1.37
1920 - 35	0.08	1.61	0.98	0.91
1936 - 51	0.04	0.54	1.27	2.07

Table 10.3. Changes in pH of soil at different depths in the Park Grass and Geescroft Wilderness Experiments (Johnston et al. 1986; Powlson & Johnston, 1994).

	Treatments and sample depths (cm)				
	Nil (natural inputs)			**N₃PKNaMg***	
EXPERIMENT AND YEAR	**0-23**	**23-46**	**46-69**	**0-23**	**23-46**
Grassland (Park Grass)					
1856	5.6	-	-	5.6	-
1876	5.4	6.3	6.5	4.3	6.3
1923	5.7	6.2	-	3.8	4.4
1959	5.2	5.3	-	3.7	4.1
1984	5.0	5.7	-	3.4	4.0
1991	4.8	5.4	5.7	3.2	3.8
Woodland (Geescroft Wilderness)					
1883	7.1	7.1	7.1	-	-
1904	6.1	6.9	7.1	-	-
1965	4.5	5.5	6.2	-	-
1983	4.2	4.6	5.7	-	-
1991	4.3	5.1	6.0	-	-

* 145 kg N/ha as ammonium sulphate, 34 kg P/ha as single superphosphate, 224 kg K/ha as potassium sulphate, 16 kg Na/ha as sodium sulphate and 11 kg Mg/ha as magnesium sulphate

Acidification has also occurred on many plots of the Park Grass Hay Experiment to which nitrogen was supplied as ammonium sulphate (Table 10.3). As it had been in grass for a century or more before the experiment began in 1856, the soil was initially slightly acid (pH 5.6). Plots receiving the largest amounts of ammonium sulphate now have a topsoil pH of 3.2 where unlimed (Powlson & Johnston 1994), which is enough to mobilise aluminium from soil clays. The nil plot has a pH of 4.8 in the surface soil (0-23cm). In the nearby Geescroft Wilderness, which was part of an arable field (pH 7.1) before setaside in 1886, the pH of the surface soil is now 4.2. So acidification without the influence of ammonium sulphate has been much quicker on Geescroft (about 3 pH units in 100 years) than on Park Grass (1 pH unit in 140 years). In the Wilderness the subsoil has also become more acid to at least 69cm depth (Table 10.3). The main reason for the different rates is probably that the tree canopy (mainly oak) now present on Geescroft Wilderness traps atmospheric pollutants more effectively than the sward of Park Grass (Johnston et al. 1986).

Broadbalk Wilderness had a more calcareous soil than Geescroft Wilderness when it was setaside in 1882, because it had been more generously chalked as an arable field. As a result it has only recently started to acidify, and the rate of accumulation of soil organic matter has been about three times faster than in Geescroft Wilderness (Jenkinson 1991). Organic matter has also accumulated faster than on the plot of Broadbalk Winter Wheat which has received 35t FYM/ha annually since 1843 (Jenkinson 1971). The pH difference between the two Wildernesses also explains why the Geescroft tree canopy is almost entirely oak, whereas that of Broadbalk Wilderness also contains ash and sycamore.

Park Grass was limed to counteract the developing soil acidity in 1883 and 1887 but without much effect. In 1903 a regime of liming half of each plot every 4 years was introduced, and in 1965 each half-plot was further divided (Warren et al. 1965); since then three of the sets of quarter-plots have received chalk calculated to maintain pH values of 5, 6 and 7, and the fourth remains unchalked with pH values now ranging from 3.2 to 5.7 depending on fertiliser treatment. Within a few years of starting the experiment Lawes and Gilbert (1859) noticed that the different combinations of nutrients were influencing the botanical composition of the sward as well as the yield of hay. Recent yields range from 1.1 t/ha dry matter (pH 3.5, 48kg N/ha as ammonium sulphate) to 8.0 t/ha (pH 7, 144kg N/ha as ammonium sulphate plus phosphorus, potassium, sodium, magnesium and silicon). On most of the plots yields decrease with decreasing pH. The nil plots have the most diverse plant assemblages (50-60 species equally distributed between grasses and dicotyledons, including legumes such as meadow vetchling and red clover), but yields are small (Thurston et al. 1976). On the unlimed plots receiving only ammonium sulphate the herbage is dominated by a limited range of acid-tolerant grasses, but where lime is added there is a wider range of grasses and some broad-leaved species. Where N is supplied as sodium nitrate, there are at least 30 species present and liming has little effect on the herbage composition.

Despite unchanging fertiliser treatments, the frequent botanical surveys of Park Grass have shown that the flora of many plots has changed slowly, often with periods of transient dominance by some species that were later displaced, or with eventual reduction of the original mixed flora to a monoculture. Some species, such as sweet vernal grass, occur on several different plots, and have become so adapted genetically to the different pH and nutrient conditions that populations from one plot will not grow satisfactorily on others, even after a period of propagation in uniform environmental conditions off the field (Snaydon & Davies 1976).

FYM was applied to two plots for a few years, but without annual tillage, as on the arable experiments, it was not adequately incorporated into the soil, so the annual applications were terminated. One organic plot now receives manures on a 4-year cycle: FYM, none, fish meal, none. Its yield (5.3-7.9t/ha) is slightly less than that of the best inorganic fertiliser treatment, but the flora is only moderately diverse with 16-25 species according to lime applications.

Residual effects of fertilisers

Soil nitrogen is subject to many rapid transformation processes, often involving losses such as denitrification to gases transferred to the atmosphere or leaching of nitrate to groundwaters, so increases resulting from application of inorganic N as nitrate or ammonium are short-lived. Some of the N in organic manures such as FYM is, however, less mobile and, as the soil organic C content increases with repeated applications, total soil N also increases so as to maintain a fairly constant C/N ratio. When FYM applications cease, the transformation of this more resistant organic N to forms of mineral N available to plants is a fairly slow and protracted process, as on the Hoosfield plot which received FYM from 1852 to 1871 only. The same is true for other nutrients, such as P and K, contained in organic manures. Because of various fixation processes in soil, inorganic forms of P and K are also stored for long periods and may be released slowly to crops, especially if no further applications are made.

The assessment of the residual values of past applications of organic manures and inorganic P and K became important after the Agricultural Holdings Act was passed in 1875. This provided compensation for outgoing tenant farmers to cover the value of improvements made during their tenancies, including the value of unexhausted residues of manures and fertilisers applied to fields. Lawes (1875) and the Royal Agricultural Society of England (Voelcker 1876) had disagreed over formulae for calculating compensation for animal foodstuff residues, and Woburn Experimental Farm was originally made available by the 9th Duke of Bedford to start experiments that would settle the matter. However, problems arose because of the yield decreases at Woburn already discussed, and also the experiments there gave no data on residues of inorganic fertilisers.

Consequently, in 1901, A D Hall (the successor to Lawes and Gilbert as Director of Rothamsted) modified an earlier experiment at Rothamsted (the Exhaustion Land) to rectify the deficiencies of the Woburn experiments. From 1856 the plots had grown wheat with variable annual dressings of N, P and K, and from 1877 to 1901 potatoes were grown and some of the plots were treated with FYM. By 1901 potato yields had declined on all plots (Johnston & Poulton 1977), possibly because of increasing nematode infestation in monoculture. From 1902 to 1939 no nutrients were added and cereals were grown to measure the residual effects of the previous treatments. These were small in the absence of fresh N. From 1940 to 1985 N alone was applied to spring barley, initially at a single rate (88 kg/ha/yr) but at four rates (0-144 kg/ha/yr) from 1976 onwards. This N increased yields, allowing the crop to take advantage of P and K residues remaining in the soil from the 1856-1901 period. The effects of these residues were initially quite large, giving 2-3 t/ha more grain than where no inorganic P, K or FYM had been applied before 1901, and even in the 1980s the difference was 0.9-1.7 t/ha, depending on the N application rate. Table 10.4 gives the mean yields for the 10 years during which N was applied at four rates; over this period the residues of artificial fertilisers increased yields by up to 1.5 t/ha/yr, and the residues from farmyard manure increased them by up to 2.3 t/ha/yr. The available soil P (soluble in 0.5 M sodium bicarbonate) at the end of this period ranged from 1-6 ppm depending on P applications in the 1856-1901 period.

Since 1986 the treatments have included N alone, N + K and N + K + three rates of P to see how quickly the amounts of soil P can be built up to a level at which they no longer limit yield. These were tested on spring barley in 1986-1990 and a test with winter wheat has continued since 1991. Responses to fresh P have been large, including those on plots given P fertiliser and FYM before 1901. However, on plots which have not received fresh P, residues from the farmyard manure applied between 1877 and 1901 have increased wheat yields from 2.5 to 6.8 t/ha (Poulton et al. 1993). Soil available P now ranges from 1 to

51 ppm, but values above 10 ppm do not increase grain yields above maxima of 5 t/ha (barley) and 8 t/ha (wheat). The latest results suggest that quite small fresh applications of P restore cereal yields on previously exhausted soils almost immediately.

Table 10.4. Effects of residues of P and K applied 1856-1901 on yields of spring barley (t/ha/yr) given different N applications in 1976-85 after no fertiliser 1902-39.

	Treatments 1856-1901		
N rate 1976-85 (kg/ha/yr)	Nil	Artificial fertilisers (1410 kg P + 5040 kg K/ha)*	Farmyard manure (1260 kg P + 3920 kg K/ha)*
0	0.9	1.7	2.2
48	1.3	2.5	3.2
96	1.4	2.8	3.7
144	1.5	3.0	3.7

* Calculated total amounts of P and K applied 1856-1901 (artificial fertilisers) and 1877-1901 (farmyard manure).

Pollutants in inorganic fertilisers and organic manures

Although fertilisers and manures are used for their beneficial effects on soil nutrient status and crop yields, concern has also been expressed recently that they may introduce pollutants into soils, which are then taken up by crops and enter the food chain. The archive of crop, soil, fertiliser and manure samples taken from the Rothamsted classicals and other field experiments provides useful evidence in these respects.

Samples from the nil plot of Broadbalk show that the cadmium (Cd) content of the soil has increased by an average of 5.4g Cd/ha/yr since 1846, though most of the increase has occurred since 1920 (Jones et al. 1987). This must have come from atmospheric deposition because the plot has never received any fertiliser or manure over this period. There has been a larger increase on the FYM plot, but the increase on the superphosphate-treated plot since 1881 has averaged only 2.6g Cd/ha/yr (Rothbaum et al. 1986), so the fertiliser applications (containing 2g Cd/ha/yr on average) have not caused any greater accumulation in the soil than that resulting from atmospheric deposition. This is probably because any Cd retained is by association with soil organic matter, which has increased very little on the plots treated with inorganic fertilisers; the remainder is lost by leaching. Grain samples from a plot given inorganic fertilisers show a small irregular increase in Cd content over the last century (Jones & Johnston 1989), but those from the FYM plot show a decrease, probably because the organic matter has retained Cd against crop uptake as well as loss by leaching.

The increase in atmospheric deposition of other pollutants, such as polynuclear aromatic hydrocarbons (PAHs), over the past century has also been studied in archived soil samples from the nil plot of Broadbalk. PAHs are carcinogenic compounds formed mainly by combustion of fossil fuels and organic wastes. The total PAH burden of the topsoil has increased fivefold over this period (Jones et al. 1989a), though some of the individual compounds have increased up to 30 times and total inputs must have been greater because some PAHs would have been lost by evaporation, decomposition or leaching to subsoil horizons. Despite this, the amounts of PAHs in grain samples from Broadbalk have decreased over the past 60 years (Jones et al. 1989b), suggesting that there is little uptake of PAHs by wheat.

The long-term effects of sewage sludge applications have been studied in samples from the Market Garden Experiment at Woburn (1942-67). The sludges used between 1942 and 1961 in this experiment came from industrial areas of London and contained quite large amounts of heavy metals, notably cadmium, zinc and copper. Even about 30 years after the last sludge application, large amounts of these remain in the soil and severely decrease the size of the total microbial population (Brookes & McGrath 1984), and limit the nitrogen-fixing ability of the rhizobial populations associated with legumes. For example, clover plants grown on plots treated with sludge and inoculated with *Rhizobium leguminosarum* var. *trifolii* produce numerous white root nodules which are ineffective in fixing nitrogen, whereas plants from FYM-treated plots produce fewer, larger pink nodules that fix nitrogen normally (McGrath et al. 1988; Giller et al. 1989).

In addition the heavy metals have toxic effects on the plants themselves. Sunflower plants grown in a ^{14}C-enriched atmosphere were 22 percent smaller (shoots + roots) in the metal-contaminated soil and total amounts of ^{14}C transferred to the soil (as roots, root exudates and soil respiration) were 30 percent less (Chander & Brookes 1991a). The decreased C inputs to the soil decreased the amount of substrate available to the soil microbial biomass, and this partly accounts for the smaller size of the total population. However, if a substrate such as maize residues is added to the metal-contaminated soil the increase in microbial biomass is only about half that in soil not treated with sludge (Chander & Brookes 1991b), so the metals also have a toxic effect on the biomass. Possibly the micro-organisms in the metal-contaminated soil divert some substrate C into energy-consuming mechanisms which protect them from the metals. Other microbially-mediated processes in soil, such as release of nitrate by mineralisation of organic matter or enhanced plant uptake of P by symbiotic association with mycorrhizae, are probably affected by the heavy metals in sewage sludge, but are less sensitive than rhizobial nitrogen fixation, at least at the levels in the plots of the Market Garden Experiment (McGrath 1994).

Discussion and conclusions

The Rothamsted Classicals and other long-term field experiments at

Rothamsted and Woburn Experimental Farm demonstrate very clearly that long-continued applications of both inorganic fertilisers and organic manures have effects on soil properties, and thereby on crop yields and semi-natural plant assemblages. By increasing supplies of major plant nutrients (N, P and K), both fertilisers and manures have beneficial effects on crop growth in the short and long term. The rapid cycling of N and losses by leaching or denitrification mean that inorganic applications of this nutrient have short-term effects (usually <1 yr), but applications of inorganic P and K and of farmyard manure can result in increases in stored soil nutrients that persist for at least several decades and may then influence crop yields if growth is stimulated by fresh inorganic N applications.

Despite widespread opinion that prolonged cereal monoculture by use of artificial fertilisers may degrade soil and lead to loss of yield, there has been little evidence for this over the century and a half of experimentation at Rothamsted. Use of ammonium sulphate as a source of N leads to soil acidification, which may have deleterious secondary effects, such as an increase in soil-borne pests and diseases, but this can be largely avoided by use of nitrate as a source of N and by regular liming to maintain a soil pH >6.5. Because of the buffering effect of soil clay, acidification is delayed for much longer periods in clayey or loamy soils than sandy. The pest and disease effects are best avoided by crop rotation, which is probably more important on sandy than on heavier soils.

Soil organic matter content is also influenced by prolonged use of fertilisers and manures. Growth of inorganically fertilised arable crops on soil previously under pasture or heavily manured leads to a decrease in soil organic matter, but this is not linear. In time the organic content stabilises at a new level in which inputs from crop residues (e.g. roots and stubble of cereals) balance losses by mineralisation and crop offtake. This new equilibrium level is likely to be much less in sandy soils than heavier loams and clays, and may weaken their structure and so increase their susceptibility to erosion. However, there is no firm evidence that the decrease in soil organic matter directly results in yield decline; early indications of this on the sandy soil at Woburn are better interpreted as a result of a parallel increase in acidity from use of ammonium sulphate, leading to greater incidence of pests and diseases.

The increase in soil organic matter resulting from repeated heavy dressings of farmyard manure is useful in sandy soils under continuous arable cultivation to improve water retention and inter-particle bonding (resistance to erosion), but has much less benefit in heavier soils. The main problem with farmyard manure is the difficulty of ensuring an adequate supply, especially since mixed farming has been replaced by intensive arable and dairy farming or beef production in well separated regions. The main alternative organic manure likely to be widely available in sufficient quantities, namely sewage sludge, can cause long-term soil degradation by heavy metal pollution, unless its composition is carefully controlled. The metal pollutants have persistent toxic effects on plants and on the soil microbial population, the most disastrous of which is to impair the nitrogen-fixing ability of leguminous plants. The Rothamsted and Woburn long-term experiments thus provide no clear evidence that either organic manures or artificial fertilisers are better in all circumstances.

The archived soil and crop samples taken from the experiments over the last 150 years show that air pollution by modern industry and vehicle exhaust emissions has caused at least as much soil degradation and water pollution as the use of fertilisers and manures. For example, acidic inputs (sulphur and nitrogen compounds) from the atmosphere have been enough to change the soil pH under the strongly scavenging influence of woodland by 3 units over the last century, whereas the acidification resulting from 135 years of atmospheric inputs to grassland plus the use of ammonium sulphate without lime for the same period was equivalent to only 2.4 pH units. Other pollutants from the atmosphere include heavy metals, such as cadmium and lead, and organic toxins including dioxins and polynuclear aromatic hydrocarbons. The archive of stored samples from the Rothamsted Classicals shows that although the amounts of these pollutants deposited on nil plots increased considerably during periods of increased industrial activity in the nineteenth and twentieth centuries, the amounts taken up by crops have increased hardly at all. They have been retained by the soil, often by association with soil organic matter, or have been slowly removed by decomposition or leaching processes. In these and other ways soil performs a useful purifying role, often protecting food supplies and potable waters from the worst influences of industrial pollution.

Bibliography

Boyd D A, Cooke G W, Dyke G V, Moffatt J R, Warren R G. 1962. The Rothamsted Ley-Arable Rotation Experiment. *Rothamsted Experimental Station Report for 1961*: 173-80

Brookes P C, McGrath S P. 1984. Effects of metal toxicity on the size of the soil microbial biomass. *Journal of Soil Science* 35: 341-46

Catt J A, Henderson I F. 1993. Rothamsted Experimental Station - 150 years of agricultural research. The longest continuous scientific experiment? *Interdisciplinary Science Reviews* 18: 365-78

Chander K, Brookes P C. 1991a. Plant inputs of carbon to metal-contaminated soil and effects on the soil microbial biomass. *Soil Biology and Biochemistry* 23: 1169-77

Chander K, Brookes P C. 1991b. Microbial biomass dynamics during the decomposition of glucose and maize in metal-contaminated and non-contaminated soils. *Soil Biology and Biochemistry* 23: 909-15

Dyke G V, George B J, Johnston A E, Poulton P R, Todd A D. 1983. The Broadbalk Wheat Experiment 1968-78: yields and plant nutrients in crops grown continuously and in rotation. *Rothamsted Experimental Station Report for 1982*, Part 2: 5-44

Garner H V. 1959. The oldest experiment in the world. *Discovery* July 1959: 296-301

Garner H V, Dyke G V. 1969. The Broadbalk yields. *Rothamsted Experimental Station Report for 1968*, Part 2: 26-46

Giller K E, McGrath S P, Hirsch P R. 1989. Absence of nitrogen fixation in clover grown on soil subject to long-term contamination with heavy metals is due to survival of only ineffective *Rhizobium*. *Soil Biology and Biochemistry* 21: 841-48

Jenkinson D S. 1971. The accumulation of organic matter in soil left uncultivated. *Rothamsted Experimental Station Report for 1970*, Part 2: 113-37

Jenkinson D S. 1977. The nitrogen economy of the Broadbalk Experiments. I. Nitrogen balance in the experiments. *Rothamsted Experimental Station Report for 1976*, Part 2: 103-9

Jenkinson D S. 1991. The Rothamsted long-term experiments: are they still of use? *Agronomy Journal* 83: 2-10

Jenkinson D S, Hart P B S, Rayner J H, Parry L C. 1987. Modelling the turnover of organic matter in long-term experiments at Rothamsted. *INTECOL Bulletin* 15: 1-8

Johnston A E. 1973. The effects of ley and arable cropping systems on the amounts of soil organic matter in the Rothamsted and Woburn Ley-Arable Experiments. *Rothamsted Experimental Station Report for 1972*, Part 2: 131-59

Johnston A E. 1975. Experiments made on Stackyard Field, Woburn, 1876-1974. I. History of the field and details of the cropping and manuring in the Continuous Wheat and Barley Experiments. *Rothamsted Experimental Station Report for 1974*, Part 2: 29-44

Johnston A E. 1986. Soil organic matter, effects on soils and crops. *Soil Use and Management* 2: 97-105

Johnston A E. 1989. The value of long-term experiments in agricultural research. In: Brown J R (ed). *Proceedings of the Sanborn Field Centennial. A celebration of 100 years of agricultural research* . University of Missouri, Columbia, 2-20

Johnston A E. 1991. Soil fertility and soil organic matter. In: Wilson W S (ed). *Advances in soil organic matter research: the impact on agriculture and the environment.* Royal Society of Chemistry, Cambridge

Johnston A E, Mattingly G E G. 1976. Experiments on the continuous growth of arable crops at Rothamsted and Woburn Experimental Stations: effects of treatments on crop yields and soil analyses and recent modifications in purpose and design. *Annales Agronomiques* 27: 927-56

Johnston A E, Penny A. 1972. The Agdell Experiment, 1848-1970. *Rothamsted Experimental Station Report for 1971*, Part 2: 38-68

Johnston A E, Poulton P R. 1977. Yields on the Exhaustion land and changes in the NPK content of the soils due to cropping and manuring, 1852-1975. *Rothamsted Experimental Station Report for 1976*, Part 2: 53-85

Johnston A E, Goulding K W T, Poulton P R. 1986. Soil acidification during more than 100 yr under permanent grassland and woodland at Rothamsted. *Soil Use and Management* 2: 3-10

Jones K C, Johnston A E. 1989. Cadmium in cereal grain and herbage from long-term experimental plots at Rothamsted, UK. *Environmental Pollution* 57: 199-216

Jones K C, Symon C J, Johnston A E. 1987. Retrospective analysis of an archived soil collection. II. Cadmium. *The Science of the Total Environment* 61: 75-89

Jones K C, Stratford J A, Waterhouse K S, Furlong E T, Giger W, Hites R A, Schaffner C, Johnston A E. 1989a. Increases in the polynuclear aromatic hydrocarbon content of an agricultural soil over the last century. *Environmental Science and Technology* 23: 95-101

Jones K C, Grimmer G, Jacob J, Johnston A E. 1989b. Changes in the polynuclear aromatic hydrocarbon content of wheat grain and pasture grassland over the last century from one site in the UK. *The Science of the Total Environment* 78: 117-30

Lawes J B. 1875. On the valuation of unexhausted manures. *Journal of the Royal Agricultural Society of England*, Series 2, 11: 8-38

Lawes J B, Gilbert J H. 1859. Reports of experiments with different manures on permanent meadow land. *Journal of the Royal Agricultural Society of England*, Series 1, 19: 552-73

McGrath S P. 1994. Effects of heavy metals from sewage sludge on soil microbes in agricultural ecosystems. In: Ross S M (ed). *Toxic metals in soil-plant systems.* Wiley & Sons, Chichester

McGrath S P, Brookes P C, Giller K E. 1988. Effects of potentially toxic metals in soil derived from past applications of sewage sludge on nitrogen fixation by *Trifolium repens* L. *Soil Biology and Biochemistry* 20: 415-24

Poulton P R, Barraclough P, Bollons H. 1993. Phosphorus nutrition in crops. *Report AFRC Institute of Arable Crops Research for 1992*: 18-19

Powlson D S, Johnston A E. 1994. Long-term field experiments – their importance in understanding sustainable land use. In: Greenland D J, Szabolcs I (eds). *Soil resilience and sustainable land use*. pp. 367-94. CAB International, Wallingford

Rothbaum H P, Goguel R-L, Johnston A E, Mattingly G E G. 1986. Cadmium accumulation in soils from long-continued applications of superphosphate. *Journal of Soil Science* 37: 99-107

Snaydon R W, Davies M S. 1976. Rapid population differentiation in a mosaic environment. IV. *Anthoxanthum odoratum* L. at sharp boundaries. *Heredity* 3: 9-25

Thurston J M, Williams E D, Johnston A E. 1976. Modern developments in an experiment on permanent grassland started in 1856: effects of fertilizers and lime on botanical composition and crop and soil analyses. *Annales Agronomiques* 27: 1043-82

Voelcker J C A. 1876. On the theoretical and practical value of purchased food and of its residue as a manure. *Journal of the Royal Agricultural Society of England*, Series 2, 12: 203-30

Warren R G, Johnston A E. 1967. Hoosfield Continuous Barley. *Rothamsted Experimental Station Report for 1966*: 320-38

Warren R G, Johnston A E, Cooke G W. 1965. Changes in the Park Grass Experiment. *Rothamsted Experimental Station Report for 1964*: 224-28

Chapter 11

The Potential and Significance
of Field System Remains

Sally M Foster and Richard Hingley

Introduction

Now, as in the past, farmers play the largest role in shaping the appearance of the countryside. Their decisions about how, where and when to develop and exploit the land's agricultural potential through time have resulted in today's 'cultural landscape'. Archaeologists employ a panoply of multi-disciplinary techniques, from simple observation to the most exhaustive and up-to-date scientific techniques, to attempt to disentangle the work of past peoples in creating our modern landscape. From this evidence we can begin to understand the complex development of society over approximately the last nine thousand years, and the inter-relationship of people with the physical features of the landscape. Herein lies the clue to much of Scotland's past – the history of its farming peoples through time. It was in the countryside – until the Industrial Revolution, urban growth, the various phases of enforced Clearance, voluntary emigration and agricultural improvements of the eighteenth and nineteenth centuries – that the vast majority of Scotland's population could be found.

The physical remains of these farming communities – their settlements and fields – survive throughout Scotland. They range from the rare evidence of activity attributed to the first Neolithic farmers (around six thousand years ago) and the more common remains of Bronze Age and Iron Age farmers (Halliday 1993), the largely 'invisible centuries' of the medieval period (Yeoman 1991; Corser 1993) to the ubiquitous remains of the post-medieval/pre-modern period (see, for example, Dixon 1993 and this volume). The later remains subsume much earlier material, probably including the elusive Early Historic and medieval evidence, which has either not been recognised (because it has been built on, cultivated, or left no physical trace) or is not recoverable on the basis of field survey alone. There has been little excavation of any Scottish medieval or later rural settlement remains (e.g. Mackay 1993, but see Fairhurst 1968; 1969), so most observations are based on survey evidence alone.

The remains often survive as upstanding features in much of the Scottish uplands. These uplands are now marginal areas, which were periodically used and, fortunately for us, abandoned at various stages in the past. Archaeological features survive especially well in the upland/marginal areas because they have not been extensively exploited in modern times until relatively recently (see illustrations in Dixon, this volume; Halliday 1993). Yet these only represent facets of a much wider landscape in which it was the lowlands that were most densely and extensively exploited. Remains of all periods survive in the low-lands, either fortuitously as undisturbed 'islands' within 'seas' of cultivation or, more commonly, under modern farms (where they usually cannot be detected) or ploughed fields (Corser 1993). Plough-truncated features can, however, be detected in favourable conditions (for instance, as crop marks or soil marks from the air; Dixon, this volume, Pl. 2.6) and may be surprisingly well pre-served. Where they are not identifiable as crop marks, these sites tend only to be located by accident during the course of archaeological assessment and excava-tion or other ground disturbance, such as ploughing (cf. Yeoman 1991, 120).

The subject of this paper is medieval or later rural settlement remains (MOLRS), and in particular the associated field systems. In order to avoid confu-sion, readers should note that archaeologists often use the term 'field remains' to embrace all physical evidence of the past, as with the concept of 'field ar-chaeology'; in this context, however, the term 'field systems' refers to land-scapes of agricultural exploitation.

Before discussing this further, we should first clarify what we mean by 'field systems'. As will be clear from other chapters in this volume (Dodgshon, Dixon, Mercer and Tipping) and the papers in Hingley (1993a), the agricultural landscape consisted of more than enclosures, cultivation remains, or associated agricultural buildings such as barns, mills or shielings. All these elements were utilised as part of a complex, ever-changing web of seasonal and annual agricul-tural activities which incorporated open areas (notably upland grazings) that may have not have left any physical imprint in the landscape (Bangor-Jones 1993).

So how in our preservation and management strategies are we to cope with these extensive field systems, particularly the problem of open areas, and the lowlands where so much evidence may remain but is not yet recognised? The problem is that while we may recognise the spatial and temporal integrity of the cultural landscape, measures for managing large-scale landscape units in their entirety rarely exist. In practical terms, even where they do (see below), we can only seek to preserve from imminent destruction what we already know to exist and consider to be important. Within longer-term strategies for protection the likely significance of unrecognised remains (particularly in the lowlands) and their undoubted research potential should also not be overlooked.

In recent years MOLRS and associated field systems have become a specific in-terest for Historic Scotland (HS), the government agency responsible for the preservation and management of the 'built heritage'. This is due to the fact that large tracts of the countryside are covered with upstanding remains which be-long to this broad period and which are, of course, vulnerable to destruction. The fields of these settlement landscapes are of fundamental importance to our understanding of the past, since they provided much of the diet of past popula-tions. These remains are important not only because of their quality and extent, but because they are the only surviving remains of Scotland's past, and of a 'silent' population which largely consisted of those without any degree of mate-rial wealth. Theirs is a largely undocumented history which archaeology (and the wide range of disciplines which this subject embraces) is uniquely placed to both identify and attempt to understand. Ironically it is the ubiquity and extensive nature of these remains, especially the associated field systems, which has led to their increasing destruction by modern land-use changes, particularly afforestation.

Threats

During the late 1980s, a system was introduced, through discussions between HS and the Forestry Commission, to minimise the risk of damage through tree planting to important archaeological sites. A system of formal notification allows an appropriate archaeologist (usually the Regional Archaeologist) to provide information and advice to the Forestry Authority (Breeze 1989; Barclay 1992). In the case of MOLRS, while it has been easy to argue a case for preserving settlement cores, in practice the field systems have fared less well. To take the case of Strathclyde Region (Swanson 1993), the Regional Archaeologist has taken a pragmatic line, to preserve well-defined and compact field systems through a process of negotiation, seeking wherever possible to preserve field banks; she has, however, found it difficult to argue for the preservation of more amorphous remains. MOLRS remains are also threatened by a range of other pressures. For instance, some ruined buildings are built into modern homes (often holiday cottages; Swanson 1993); this tends to pose less of a direct threat to the surrounding field systems, although the effect of associated developments, such as new roads, may be unsympathetic to landscape preservation. Other MOLRS sites, with their field systems, are threatened by road building (e.g. McCullagh 1991) and by various other forms of development (for instance agricultural improvement, pipeline construction, etc.).

Historic Scotland's approach

HS recognised that a range of archaeological resource managers in Scotland were facing a problem in providing an archaeological response to development proposals involving MOLRS sites. As a result, a seminar was organised in November 1991 to discuss relevant issues. The papers presented at the seminar, along with the discussion, has now been published as a Historic Scotland Occasional Publication (Hingley 1993a).

The discussion that occurred at the seminar led to the following observations:

> It is important that defensible criteria for the selection of field systems for preservation are developed for those settlements where the archaeologist wishes to protect the whole settlement and agricultural complex. It is also important that adequate standards are developed to record archaeological features where preservation is not possible.

> . . . heightened awareness of the need for protection will come partly through a recognition that these sites are not only of interest to the historian and the archaeologist, but are also a vital part of our local and national consciousness (Hingley 1993: 58, 61).

With these considerations in mind, HS set up an Advisory Group in 1992 which has since met regularly to discuss how the wider issues of MOLRS might be progressed (its aims are outlined in Hingley 1993a: 66-7.) HS also commissioned the compilation of a Secondary Sources Index. This comprised the creation of a register of interested individuals and institutions, and an index of secondary sources (primarily unpublished research) on MOLRS. In addition a study was undertaken of the documentary evidence available for Kyle and Carrick. This was a pilot project to see how/if documentary sources might practically aid decisions about which MOLRS remains should be selected for preservation. Kyle and Carrick was selected because it was an area with numerous well-preserved and extensive remains (listed by the Royal Commission on the Ancient and Historical Monuments of Scotland – RCAHMS) which continue to be actively threatened. Both of these reports are available for consultation in the National Monuments Record of Scotland (Clapham 1993; Watson 1993).

It soon became clear in the case of the Kyle and Carrick study that although there was considerable documentary evidence for medieval and later settlement, and much of this did refer to named sites, these could not necessarily be identified on the ground nor conclusively associated with any of the known remains. The main value of the report was therefore recognised as the need to consider MOLRS on a wider canvas and provide a wider historical context into which such remains might be placed. Furthermore, this work was expensive and time-consuming and Historic Scotland, with its national remit, could neither afford nor wait for such work to be undertaken throughout Scotland as a whole, particularly when the pilot study had shown only limited potential to fulfil HS's requirements directly. Historic Scotland was also concerned that it was not the appropriate body to undertake such detailed documentary research as Kyle and Carrick entailed. Furthermore, it is the physical remains which are under pressure from external threats, not the historical sources: we therefore needed an approach to the preservation of these remains which recognised the historical dimension, but at the same time acknowledged the unique potential contribution of the physical remains themselves.

Assessment of importance

HS's provisional view is that three considerations relating to the preservation of MOLRS remains may be identified; sites/landscapes may be defined to be of particular importance for the following reasons:

1. The potential of a site to address important academic questions;
2. The presence of well-defined physical characteristics and other considerations which are used to assess whether field remains are of importance: survival/condition; period; group value; rarity; situation; multi-period/single period; fragility/vulnerability (see Hingley 1993 for further details);
3. Important historical associations/place in the consciousness of the modern population.

These are obviously no more than general criteria, which could be applied at local, regional or national level. For scheduling, however – i.e. legal protection under the *Ancient Monuments and Archaeological Areas Act 1979* (see Hingley 1993 for further details) – a case would have to be made that the sites/remains were of national importance. The three considerations mentioned above are obviously not mutually exclusive, but the first point is the most important, because not only does it subsume archaeological, environmental, historical and other types of potential, but it also avoids the danger of producing criteria so vague that they might encompass all sites. It is the research issues which will provide the more specific considerations, and it is the definition of these which will now hopefully engage all those interested in this subject. So, whilst some sites might be preserved for predominantly historical considerations (Arichonan, mid Argyll, is a good example; see RCAHMS 1993: 454-7), in most cases the justification for preservation would be an informed assessment of the potential of the physical remains to address key issues, defined from a multi-disciplinary perspective. In order to preserve a representative sample of these remains we therefore need to know more about the character of what survives on the ground, and a better understanding of the types of questions such remains can answer.

Research potential

To return to the field systems, what sort of questions might they have the potential to answer? It was because we still know so little about medieval or later field systems that HS organised its second seminar on this topic. Our aims were to obtain an overview about what was surviving and where (Fenton and Dixon, this volume) and to consider the issue of minimum standards for recording. Yet perhaps the most important question was an assessment of the archaeological potential of the field systems. Could we be certain that they were worth preserving at all (Dodgshon and Hall, this volume)?

Despite the fact that very little work has yet been undertaken on this subject, it became clear that, given use of the appropriate techniques, field systems *do* have enormous potential, even if in some cases the scientific work is at an early state or may currently be beyond the reach of the public purse. This potential lies both in the field banks (which tend to preserve earlier features from the erosive effects of surrounding cultivation) *and* the interior of the fields, although it was acknowledged that the greater potential usually lies in the fields closer to the settlement cores. It was clear from Dodgshon's work (*op. cit.*) that in order to understand nutrient transfer, for example, whole townships/related settlements would need to be preserved. It was also clear that their value as landscape features was widely appreciated, notably by Scottish Natural Heritage. Clearly much more work is needed, and it is especially gratifying to note that a relevant project has recently been set up jointly by Stirling University and AOC (Scotland) Ltd to examine the relationship between field morphology and land-use through the examination of the soils (Donald Davidson, *pers comm*).

Preservation options

Increasing numbers of field systems (and associated settlements) continue to be preserved and better managed through a variety of means (summarised in Hingley 1993; see also Macinnes 1993 for wider discussion of options for management in the UK). In the case of afforestation, most recent planting proposals affecting field systems have left extensive areas unplanted (*pers comm*. Tim Yarnell). As archaeologists, it is particularly rewarding to see how many field systems are now being protected through the participation of farmers in the Environmentally Sensitive Areas scheme, which extends to nineteen percent of the country. Remains ranging in importance from local to national significance can all receive protection from potentially damaging farming practices. Additionally, some of the farmers are claiming an optional payment for managing these landscapes in a manner which will enhance their long-term preservation (such as removal and control of bracken or other harmful vegetation and/or introducing suitable grazing regimes in order to control this). In talking to the farmers about these sites the agricultural and archaeological advisers are also raising awareness of the significance of these remains, as well as finding out about sites they (and we) did not know even existed! Some of the farmers are surprised to find that these remains are considered to be important. Recognition of their importance has come late to archaeologists too; it is only in more recent years that we have begun to fully record these later remains, or that the Ordnance Survey has been persuaded to publish as antiquities remains later than 1714 (Stevenson 1993). As a result, it is only in areas where the RCAHMS has recently undertaken a field survey that we have more complete and detailed coverage of the field system remains (see below).

Looking to the future, non-archaeological forms of landscape designation (such as Natural Heritage Areas) will probably offer the best opportunities for the preservation of these wider landscapes. HS will liaise closely with other interested parties, particularly Scottish Natural Heritage, who have a wider remit for countryside management, education and recreation than afforded by the legislative framework within which HS operates.

RECORDING OF FIELD SYSTEMS

Whilst we may hope to see, in collaboration with the farming community, further preservation of many of these diverse remains, the fact is that many MOLRS field systems will continue to be threatened by development. What then should we be doing to record them? In fact it is not yet easy to answer this question; only when we have a firmer idea of the range of questions these remains can potentially answer, can we identify appropriate levels of recording.

Current recording of MOLRS field systems

Current practice falls into two main categories: research work undertaken by universities, units and local societies; and strategic work, both short- and long-term, carried out by, or at the instigation of, central and local government bodies. The latter includes aerial photography and ground survey by the RCAHMS, pre-afforestation surveys commissioned by HS to aid the Regional Archaeologists, desk-based assessments in advance of developments, and subsequent fieldwork, either evaluation or excavation.

This work mainly takes the form of survey: that is, the production of maps and a photographic record; the majority of this is undertaken by the RCAHMS. Its National Archaeological Survey (NAS) and Afforestable Land Survey (ALS) teams record such remains in the course of their routine fieldwork (see RCAHMS 1990; 1993a; 1993b; 1994). *North-east Perth* is the most extensive landscape survey of medieval and later field systems yet published in Scotland. Current ALS practice is to map field banks at basic mapping scales (1:10,000 or 1:2,500). It is usually only possible to record a sample of the rigs, depending upon time constraints. Frequent use is made of existing vertical aerial photographs, blown up to a suitable scale, to supplement plans, and this information is transferred to the 1:10,000 maps sheet overlays in the National Monuments Record for Scotland (NMRS). Unusual details, such as patterns and relationships in the cultivation remains, will be described in the text if they are not planned in detail. The RCAHMS is starting to define different types of rig: for example, broad, narrow, cord and lazy beds, in order to distinguish between plough and spade cultivation (if possible) and to record the size and shape as well as relationships to other features. Ground photography has limited potential in recording field systems, but the RCAHMS aerial photography team does record field system remains in the course of its flying. The RCAHMS undertakes general but essentially limited documentary research, the relevant findings of which are incorporated in the Record. Documentary evidence relating to specific field system remains is rarely encountered.

HS's specification for the pre-afforestation surveys it commissions did not originally contain specific guidelines about the level of detail to apply to the recording of field system remains; but it is now accepted practice that all field boundaries will be recorded, the direction of the rig will be noted on the plans, and the spacing will be described in the text. Primary documentary research is not part of the standard brief.

Desk-based assessments are now a frequent component in the lives of archaeological resource managers. These are usually prepared for Environmental Assessments – usually large-scale developments – often for linear features such as pipelines or roads. The generally adopted procedure is to record the presence of field system remains and to assess the significance of the surviving remains in order to devise a mitigation strategy. This usually consists of a written description which may include measurements of rig width. Little excavation on medieval or later field system remains has been undertaken in the course of such development work since it is usually possible to avoid field banks, and developments often only impinge on a minor part of a more extensive site. If a case is made, field banks can be preserved from planting, and the move away from ploughing as a form of ground preparation is helpful in this respect.

Such information is used by HS and the Regional and Island Archaeologists to make decisions about what should be preserved and how it should be managed. In the absence of specific research agendas and detailed recording guidelines for MOLRS remains in different parts of the country, it is difficult to be sure whether the type of information collected is consistent and forms a firm basis to make informed decisions.

Decisions regarding which field systems to preserve will usually not be isolated from the nature, quality and significance of associated settlement remains. Although there are circumstances where field system remains might be preserved on landscape grounds, it is usually only possible to make a case for preservation on the basis of strict archaeological significance. In these circumstances, we need to know what survives, its extent, nature of preservation and likely potential for the recovery of archaeological and palaeoenvironmental information. We need to ensure that our levels of recording can address these questions adequately (if indeed they are the right questions to ask).

Recording in advance of destruction

Making a final record of a site in advance of its destruction should be a totally different exercise from undertaking a general survey. We need to know what sort of information we are losing – particularly in the case of afforestation – and what, if anything, should we be doing to record it before its destruction? Yet recording of these lost areas has tended to exclude, for example: an adequate modern photographic record of the field system remains in their landscape setting (and we cannot always rely on RAF photographs or other aerial surveys for this information); detailed recording of the rigging, specifically profiles of cultivation remains; and environmental data of *any* form, or dating evidence. This last point in particular is inconsistent with practice on other forms of archaeological remains, including prehistoric field systems, where they *are* adequately recorded in advance of destruction, usually as an integral part of development.

It can be argued that the archaeological remains within any one field may be repetitive, and that it is therefore not necessary to record fully the disturbance of only a part of it, but this too is inconsistent with policies for other types of extensive monuments. Field systems may be more numerous than other types, but individual systems have the potential to provide unique, localised information. Nor should we forget the truncated features, perhaps of earlier periods (particularly of timber), which lie throughout them. So, how much of any field system is it acceptable to lose without further record? What type and scale of sampling strategy should be undertaken? We know the potential of such remains, and the techniques which can be employed to tackle them; but when should we insist that they are implemented, and at whose expense?

The current position is that the forestry industry does not pay for archaeological survey in advance of planting (Barclay 1992). When the notification system was set up in 1988, it was considered that it would be unfair to expect some applicants for grant to pay while others did not have to bear any such cost, as the need for a new survey would depend solely on when the most recent archaeological fieldwork had occurred. Funds were made available to enable HS to undertake surveys immediately in advance of forestry. RCAHMS also received funds to establish its Afforestable Land Survey team which undertakes strategic survey in those areas of potential afforestation, where the archaeological record is inadequate. Since 1988 the situation in relation to general planning matters has changed and it is now a normal expectation that the developer will meet the costs of environmental works necessitated by their development (NPPG 5; PAN 42), but the forestry arrangements remain in place. It has been suggested that developer funding of archaeological survey and mitigation in advance of forestry planting might make private planting uneconomical (*BAN* 1994).

Conclusions

Medieval or later field systems (in their widest sense) contribute not only to the texture and visual amenity of our modern 'cultural landscape' but, as many of the accompanying papers demonstrate, they have the unique potential to tell us how farmers have used the land over hundreds and thousands of years. This information may not be easily gleaned from the remains, nor is it likely to come cheaply, but unless we ensure their preservation, how will we able to ask questions of our past in the future? These remains are not just of relevance to archaeologists and historians as a future repository of data, but to all those interested in the history, amenity and educational value of the countryside as a whole. Landowners and their tenants are the guardians of archaeological features in the countryside, and we must continue to raise their awareness of the significance of these remains, and work in co-operation with them to secure the long-term preservation of this irreplaceable resource.

Bibliography

Barclay G J. 1992. Forestry and Archaeology in Scotland. *Scottish Forestry* 46: 27-47.
BAN. 1994. *British Archaeological News* New Series 12: 1.
Breeze D. 1989. Forestry and Monuments: The Role of SDD HBM for Antiquities (Part 2). In: Proudfoot E V W (ed.). *Our Vanishing Heritage. Forestry and Archaeology.* Council for Scottish Archaeology, Edinburgh: 20-21.
Clapham P. 1993. *Secondary Sources Register and Register of Individuals and institutions interested in MOLRS.* Unpublished report, Historic Scotland.
Corser P. 1993. Pre-improvement settlement and cultivation remains in eastern Scotland. In: Hingley R (ed.): 15-23.
Dixon P. 1993. A review of the archaeology of rural medieval and post-medieval settlement in highland Scotland. In: Hingley R (ed.): 24-35.
Fairhurst H. 1968. Rosal: a Deserted Township in Strath Naver, Sutherland. *Proc Soc Antiq Scot* 101: 135-69.
Fairhurst H. 1969. The Deserted Settlement at Lix, West Perthshire. *Proc Soc Antiq Scot* 101: 160-99.
Halliday S P. 1993. Marginal agriculture in Scotland. In: Smout T C (ed). *Scotland since prehistory. Natural change and human impact.* Scottish Cultural Press, Aberdeen: 64-78.
Hingley R. 1993. Past, current and future preservation and management options. In: Hingley R (ed.): 52-61.
Hingley R (ed.). 1993a. *Medieval or later rural settlement in Scotland: management and preservation.* Historic Scotland, Ancient Monuments Division, Occasional Paper 1, Edinburgh. [available from Historic Scotland, Longmore House, Salisbury Place, Edinburgh EH9 1SH, £9 including postage]
Macinnes L. 1993. Archaeology as Land Use. In: Hunter J, Ralston I. *Archaeological Resource Management in the UK. An Introduction.* Alan Sutton/Institute of Field Archaeologists, Stroud.

Mackay D. 1993. Scottish Rural Highland Settlement: preserving a people's past. In: Hingley R (ed.): 43-51.

McCullagh R. 1991. Lairg. In: *Discovery and Excavation in Scotland* 1991, 46-8.

NPPG 5. 1993. *National Planning Policy Guidance No. 5: Archaeology and Planning.* The Scottish Office, Environment Department.

PAN 42. 1993. *Planning Advice Note No 42: Archaeology - the planning process and Scheduled Monument procedures.* The Scottish Office, Environment Department.

RCAHMS. 1990. *North-east Perth. An Archaeological Landscape.* HMSO, Edinburgh.

RCAHMS. 1993a. *Waternish, Skye and Lochalsh District, Highland Region - An archaeological Survey.* RCAHMS, Edinburgh.

RCAHMS. 1993b. *Strath of Kildonan: An archaeological survey.* RCAHMS, Edinburgh.

RCAHMS. 1994. *Glenesslin, Nithsdale: An archaeological survey.* RCAHMS, Edinburgh.

Stevenson J B. 1993. Introduction to the RCAHMS contribution. In: R Hingley (ed.): 11-12.

Swanson C. 1993. The need for a management and preservation strategy. In: Hingley R (ed.): 1-3.

Watson F. 1993. *Report on the survey of documentary sources for Medieval or Later Rural Settlement in Kyle and Carrick District.* Unpublished report, Historic Scotland.

Yeoman P. 1991. Medieval rural settlement: the invisible centuries. In: Hanson W S, Slater E A (eds.). *Scottish Archaeology: new perceptions.* Aberdeen University Press: 112-28.

Acknowledgements

We are grateful to Dr David Breeze, Professor Donald Davidson, Dr Lesley Macinnes, Gordon Barclay, Dr Piers Dixon, Strat Halliday (and RCAHMS colleagues) and Tim Yarnell for their comments on various drafts of this paper. We also would like to take this opportunity to thank the present and past members of the MOLRS Advisory Group who have given freely and generously of their time, and whose advice continues to be greatly appreciated: Dr Malcolm Bangor-Jones; Professor Donald Davidson; Dr Piers Dixon; Professor Tom Devine; John Hume; Dr Chris MacGregor; Professor Allan Macinnes; Aonghus MacKechnie; Dr Alex Morrison; Dr John Shaw; Professor Chris Smout; Dr Bruce Walker and Tim Yarnell.

Chapter 12

The Soil Resource and Problems Today: An Ecologist's Perspective

John Miles

Introduction

This chapter introduces the nature of soil, then briefly reviews the main problems facing soils today, with particular reference to Scotland. These include erosion, acidification, pollution by heavy metals, radionuclides, excess nitrogen and pesticides, soil burial by built developments, plus the potential or even likelihood of it responding to future changes in climate resulting from the dramatic increases in greenhouse gas concentrations in the atmosphere. The views expressed are the author's and should not be taken as those of the Scottish Office.

Soil is the surface layer of the Earth's crust in which most vascular plants root and from which they get water and mineral nutrients. It is a remarkable mixture of weathered rock particles and debris, decaying organic matter, and a wealth of soil-dwelling organisms. Soil has been likened to a living tissue – the planet's skin – in which the soil water acts like blood as it moves through a network of pores, carrying dissolved minerals. It is commonly branded by millennia of human use and abuse and at least locally contains much of the remaining muck, junk and wreckage which constitutes our archaeological and industrial heritage.

Many city dwellers in industrialised societies probably think of soil only as dust and dubs, yet it still supports most of the world's food production. It is also a reservoir of biodiversity. Most temperate terrestrial ecosystems probably contain as many or more species below ground as above ground, including microorganisms that have yielded pharmacologically vital antibiotics. Soils are fascinating to study, being complex, dynamic, heterogeneous, partly-living systems which are an integral part of all terrestrial ecosystems. Soil is naturally and continuously being lost and formed; lost by erosion, solution of minerals and oxidation of organic matter, and formed by weathering of subsoil, the incorporation of organic matter and the redeposition of material eroded from elsewhere.

The concept of soil quality can only be defined by reference to good (or desirable) and poor (or undesirable) states. Good quality soil has variously been thought of as that which has unimpaired natural processes and properties (including possession of a diverse fauna and flora), which permits normal vegetation growth and development on any particular soil, which is not a risk to animal or human health, and which does not contaminate air or water. In the context of pollution, soil quality is often defined in terms of the amounts of given pollutants which can be received without causing damage to particular processes or functions. To farmers and foresters, soil quality relates in particular to its productive capacity (including its tractability and trafficability under cultivation) – the farmer's concept of being 'in good heart'. Soil quality is thus the resultant of the effects of stresses caused by natural forces and human activities. Reduction of soil quality as a result of human activity can conveniently be thought of as degradation, contrasting with reductions of quality caused by natural processes, such as weathering and leaching, which simply represent soil development to 'old age'.

The variety of soils found in Scotland reflect especially its diverse geology. The prevalence of acid rocks as soil parent materials and the generally cool wet oceanic climate have resulted in a predominance of sandy, rather leached and acid soils, often with peat or peaty cappings, of relatively low intrinsic fertility. Although about one-third of Scotland lies below 300m, the altitude around which intensive agriculture stops, the prevalence of intrinsically poor soils is indicated by land capability for agriculture assessments (Macaulay Institute for Soil Research 1982). These predict that only 1.4% of Scotland is capable of growing a wide range of crops, less than five percent of the climatically suitable land.

Erosion

Erosion of soil particles by wind and water is a natural geomorphological process which occurs particularly when the vegetation cover is incomplete or disrupted (Thornes 1990). In most places at most times, rates of erosion are very low, but pulses of greatly enhanced erosion in the past, associated with the glacial cycles, resulted in massive translocation of soil, often on a continental scale, producing in many parts of the world the deep and fertile loess soils. However, increasing human interference with vegetation and land has also in many parts of the world produced problem – and sometimes catastrophic – levels of erosion. In general, erosion probably only matters when it begins to exceed rates of formation of new soil by weathering. The latter in Britain is in the order of 1 tonne/hectare/year, or about 0.03-0.05mm per year (Morgan 1987).

Soil erosion never used to be thought of as a British problem. Thus Symon's classic (1959) account of Scottish farming mentions the shifting dunes of Culbin but is otherwise silent on soil erosion except for a passing reference to it in the American prairies. However, increasing intensification of agriculture during recent decades has rather changed this earlier picture, and by 1987 it was estimated that 15-25 million hectares of land within the European Union were threatened by erosion (Institute of Terrestrial Ecology and Soil Survey and Land Research Centre 1989), including over a third (i.e. more than 2 million hectares) of the arable land in England and Wales. While no comparable estimates appear to exist for Scotland, locally severe wind erosion is endemic on cultivated soils derived from the Upper Old Red Sandstone around the Moray Firth. Also erosion incidents (e.g. the sudden appearance of obvious rills and small gullies in fields) have increased greatly in the east of Scotland as a result of a partial switch to winter cereal production (Spiers & Frost 1985), with a consequential increase in amount of bare ground exposed during autumn and winter. Interestingly, the same increase in extent of winter cereals in England

and Wales has led if anything to reduced soil erosion by water, apparently reflecting the earlier cultivation and sowing times possible in a more favourable climate for crop growth.

Wind erosion of soils in Britain is particularly apparent in the drier eastern areas, in sandy soils with low clay and organic matter contents, and in drained and cultivated lowland peats. Although large areas of Britain's arable soils are potentially at risk from wind erosion, this only appears to be appreciable locally or in particularly dry years. While lowland peats become susceptible to wind erosion after draining and cultivating, the major losses from these are from irreversible shrinkage and oxidation after drying. Thus the extent of the very fertile fen peatlands of East Anglia has shrunk by some 90 percent during the past 300 years, with much of the remaining area now having less than 1 metre depth of peat (Burton & Hodgson 1987).

Sandy soils and peat are also more susceptible to erosion by water. This is locally severe in lowland arable areas and in the uplands. In the lowlands, where estimates of natural erosion rates are 0.5-1.0 tonnes per hectare per year, a 5-year monitoring programme in England and Wales suggested that about five percent of the land monitored has significantly enhanced erosion in most years, rising to about 15 percent in some years, with erosion rates in excess of 15 tonnes per ha in some silty soils, and in excess of 45 tonnes per ha in some sandy soils (Evans et al. 1988). Losses may be similar in parts of east Scotland (Frost & Spiers 1984). However, the problems are recognised officially, and recent guidance should lead to reduced rates of erosion in the future (Ministry of Agriculture, Food and Fisheries 1993a).

Instances of severe erosion seem more frequent in the uplands than in the lowlands, and erosion rates may have increased in recent centuries (Innes 1983; Ballantyne 1991). Blanket peat erosion is so widespread that an eroded peat category is shown on soil maps (Macaulay Institute for Soil Research 1982). While there have been many local studies of soil erosion in the Scottish uplands (e.g. Scott 1993), no national figures are as yet available. However, at the time of writing Scottish Natural Heritage has commissioned work to begin to quantify the scale and extent of upland soil erosion and to try to assess recent trends. The causes of upland erosion are unclear, but deforestation, trampling by sheep and red deer on particularly sensitive soils and on peatlands, and perhaps burning in particular, have probably contributed. Erosion as a result of human trampling does occur, but is of very localised importance.

Ploughing for upland afforestation can cause marked pulses of erosion by water. While these losses probably do not have a significant effect on the fertility of the remaining soils (which usually are of low intrinsic productivity), appreciable amounts of sediment have often been deposited in streams and rivers (Soutar 1989). In some cases, sediments have blanketed the gravel stream beds, either preventing spawning by salmon or the successful development of the eggs and parr, and also reducing the availability of invertebrate food for the fish (Nature Conservancy Council 1990; Salmon Advisory Committee 1991), though current guidelines (Forestry Commission 1993) should minimise such incidents. However, once a tree cover is established, soil erosion rates are similar to natural rates in the lowlands, though tree felling can result in minor pulses of erosion (Moffat 1988).

Losses from building and mineral workings

Urban and industrial developments, building new roads, mineral extraction and sites for tipping waste variously bury or destroy the associated soils, though nowadays topsoil is often removed and used to rehabilitate derelict industrial sites. While about ten percent of the UK is classed as urban (Department of the Environment 1992), the figure for Scotland is only about 2.4% (Macaulay Land Use Research Institute 1993). Soils in Scotland lost to built developments and mineral workings during the 1980s amounted to about 0.14-0.17% of Scotland (Scottish Office 1991), and there were some compensatory gains from the rehabilitation of industrial areas. Losses to development are therefore small, and not irreversible, as the soils of developed areas can fairly easily (though usually not cheaply) be reclaimed for productive use, except where severe pollution has occurred.

Pollution

For convenience, elements and compounds which can contaminate soils are divided here into six classes: heavy metals, radionuclides, pesticides, hydrocarbons, nitrates and microbiological. Even though some heavy metals are essential trace elements, they are generally toxic to plants and animals at higher concentrations, though toxicity varies with the molecular form (Moriarty 1988). The relatively strong charges on the ions of most heavy metals mean that they are strongly retained in the soil, where they tend to accumulate (Berrow & Webber 1972). Heavy metals in soils are taken up by plants, can affect vegetation and crop growth (Vergnano & Hunter 1952; Goodman et al. 1973) and can be a risk to human and animal health if entering water supplies or food chains (Hutchinson & Meena 1987). Heavy metal pollution can arise naturally because the elements were present in the soil parent materials (Thornton 1983), though concentrations above trace level seem pretty well confined in Scotland to ultrabasic soils (Proctor 1971), of which few areas are in cultivation. In contrast, depositions of heavy metals from the atmosphere are ubiquitous (Cawse 1980; Purves 1985). They get into the atmosphere naturally from dust, sea spray, volcanic activity and vapour-phase outgassing of relatively volatile elements. However, heavy metal depositions only very locally become chronic (e.g. around smelters) and there are no such sites that I know of in Scotland. Nevertheless, many peaty soils bordering the industrial belt of Scotland's Midland Valley have relatively high lead contents (Paterson 1989), while even peaty soils much more remote from industrial activity have enhanced lead concentrations (Bacon 1990, 1991). Much of this lead has probably come from vehicle emissions.

A potentially more important source of heavy metal contamination of soil is from the disposal of sewage sludge by spreading on agricultural and forest land (Davies et al. 1983) though its use on the former is now controlled (Department of the Environment 1989) and its effects are still under review (Ministry of Agriculture, Food and Fisheries 1993b). While concentrations of heavy metals in sewage sludge are generally measurable in parts per million, sometimes concentrations rise to be detectable in parts of a thousand. The potential for heavy-metal contamination of soils from sewage sludge will rise as the Government implements its commitment to stop dumping sewage sludge in the North Sea by the end of 1998 and alternative sources of disposal have to be found. Interestingly, cosmetics were formerly a notable source of heavy metals away from industrial areas, especially of lead. A decade ago the problem was restricted to certain cosmetics of Asian origin but was still considered serious enough to be the subject of an official warning (Royal Commission on Environmental Pollution 1983).

Sewage sludge disposal is controllable, but any resulting heavy metal contamination of soils is persistent and not readily susceptible to amelioration. Similarly, heavy metal contamination of land as a result of industrial activities, including the disposal of contaminated waste, poses difficult restoration problems (Bridges 1988; Smith 1988; Wolf et al. 1988). Happily, the areas of land in Scotland seriously contaminated with heavy metals or other toxic substances seem to be very limited, and atmospheric concentrations are falling (Scottish Office 1991; Department of the Environment 1992).

Contamination of soils with radionuclides as a result of depositions from the atmosphere is thankfully a decreasing problem with abatement of nuclear weapons testing. Such pollution is avoidable, but if pollution occurs as a result of carelessness – eg the accident in 1986 at Chernobyl – the consequential contamination poses severe restoration problems because of the relatively long half-lives of many of the isotopes involved. Thus the accident at Chernobyl resulted in particularly high depositions of radioactive caesium-137 in many upland areas in Britain. Caesium-137 behaves very similarly to potassium, so that in mineral soils it rapidly becomes fixed, but in highly organic and peaty soils it cycles quite rapidly between soil and plants, leading to relatively persistent contamination of the vegetation on offer to grazing sheep and other animals (Horrill et al. 1989; Cheshire et al. 1991). As a result, it was necessary to control the sale and movement of ewes and lambs from many upland farms with organic soils contaminated with caesium-137 after the Chernobyl accident.

The concerns of the 1950s and 1960s about the toxic effects of relatively persistent organochlorine pesticides decreased when most of these substances were banned in the 1970s. The pesticides and herbicides in widespread use today have increasingly been tailored specifically to affect only the target organisms and also to biodegrade much more quickly in the soil. Thus the affects of pesticides after application at manufacturers' recommended rates on biological processes in the soil seem to be relatively limited (Domsch 1984) and I am not aware of any widespread problems of accumulation in Britain (though there are of course significant affects to above-ground invertebrate faunas). There is however significant use of pesticides and herbicides in non-agricultural situations, in particular along the sides of roads and railways. The main problem with pesticides now is the risk of drinking-water supplies becoming significantly contaminated.

Acute contamination of soil by hydrocarbons occurs in a few parts of the world at natural oil seeps (e.g. McCown et al. 1973), but chronic contamination is otherwise restricted to certain industrial and commercial sites, such as oil refineries, gas works and petrol filling stations. The main concern over hydrocarbons is that some, particularly certain of the polynuclear aromatics (PAHs) such as benzo[*a*]pyrene, are toxic and can cause cancers and mutations (International Agency for Research on Cancer 1973, 1979, 1983); the first clue about this seems to have been observations in 1775 that scrotal cancer in chimney sweeps resulted from exposure to soot (cited in Sims & Overcash 1983). However, as well as occurring in fossil fuels, PAHs and other hydrocarbons occur naturally by synthesis in many plants and certain animals, and as combustion products in natural forest and grassland fires caused by lightning and volcanic activity (Blumer 1976; Edwards 1983). It is therefore not surprising that micro-organisms capable of degrading hydrocarbons, including PAHs, are widely distributed in soil and in freshwater and marine sediments, that microbial communities adapt to hydrocarbon pollution and increase biodegradation rates, and that some hydrocarbons biodegrade in weeks and months (Atlas 1981; Atlas & Bartha 1992; Leahy & Colwell 1990; Wilson & Jones 1993) though higher molecular weight PAHs are more refractory and may take years to biodegrade (Wild et al. 1991).

PAHs are ubiquitous in soils and sediments worldwide, mainly reflecting a 'rain' transported in the troposphere from natural and anthropogenic combustion processes. A recent study in Wales, which avoided land near point sources of pollution, found PAH concentrations in the soil varying from about 100 to 55,000 parts per billion (ppb), averaging about 300ppb in remoter rural areas (Jones et al. 1989). Even soils in remote Swiss alps can have 4000-6000ppb (Blumer et al. 1977). Current official advice for garden and allotment soils in which crops are to be grown and where the soil may be handled and children play, is that 500,000 or more ppb of hydrocarbons constitutes contamination (Interdepartmental Committee on the Redevelopment of Contaminated Land 1987), and even such high values are unlikely to affect plant growth. These figures put into perspective the PAH levels of up to 107ppb found in soils in south Shetland in 1993 just a few weeks after a visible smear of crude oil from the tanker MV *Braer* was blown on land in sea spray (Ecological Steering Group 1994).

High levels of nitrates in soils, whether derived from inorganic fertilisers or from the disposal of animal slurry and manures, or other sources, pose even less problem for soils as such than pesticides. However they can again pose problems because of leaching into ground waters which are subsequently tapped for drinking water. Happily there is no evidence of nitrate pollution of ground waters in Scotland other than very locally and there is virtually full compliance in Scotland with the nitrates standard of the 1980 EC Drinking Water Directive (Department of the Environment 1992).

Soil is not biologically 'safe', but the only widespread risk to human health – of contracting tetanus – is routinely overcome by prophylactic injections. Dormant anthrax spores can also persist in the soil (Van Ness 1971), and probably still do so very locally in livestock farming areas given that anthrax was formerly not uncommon and that dead beasts were disposed of by burial! The risk to human health is however minute, though there used to be a more appreciable risk on Gruinard Island which was contaminated during biological warfare trials in 1942-43. The island was however successfully decontaminated in 1986 (Miles et al. 1988).

Soil compaction

Unlike many forms of pollution, soil compaction is not a new problem. Most people have probably noticed the ponding of surface water in many fields just inside the gates where there is heavy trampling by hooved animals and farm equipment. Similarly, many an owner of a newly-built house will have been dismayed by the often rock-like consistency of the surrounding soil. However, concerns about farmland grew and became more widespread in the 1960s, and reductions in crop growth were blamed on compaction caused by cultivating soils, particularly clay soils when they were too wet. This problem has been quite well researched, and wheeled vehicles and animal hooves appear to be the main cause of compaction, rather than the soil condition when cultivated. Problems of management-induced compaction are however largely restricted to clay soils, and although they can occur elsewhere they are generally more local (eg just inside field gates) and are more readily corrected by adding lime and organic matter to improve the soil structure and hence drainage. Scotland has few clay-rich soils, so severe soil compaction is a purely local phenomenon. Nevertheless, as well as deleteriously affecting plant growth, severe soil compaction is generally unfavourable to the soil fauna and flora (Institute of Terrestrial Ecology 1989).

Acidification

All soils in Britain naturally tend to become more acid with time because of the progressive leaching out of base cations in Britain's relatively high rainfall oceanic climate. This phenomenon is why farmers put so much emphasis on liming soils because there is a very large decrease in rates of nutrient cycling, and particularly in nitrogen mineralisation, at acidity levels below pH 5-5.5, with detrimental effects on the growth of all agricultural crop plants. However, natural rates of soil acidification can be increased by several land management activities, including soil drainage and fertiliser use, and also as a result of the deposition of acid pollutants from the atmosphere, the last being a particular concern since the 1970s. Acidification resulting from soil drainage and fertiliser use are relatively local phenomena which can be controlled by liming. Here I discuss only acidification resulting from (1) atmospheric pollutants, and (2) changing the vegetation cover.

Only soils with a low buffering capacity (ie a limited ability to neutralise acids) are sensitive to acidification. Thus soils with appreciable contents of carbonates or high clay contents are insensitive to acidification over centuries, while sandy soils low in clay are the most sensitive. While it was formerly thought that high organic matter contents exerted considerable buffering capacity, this appears to have been over-estimated as there is now good evidence that even peats can acidify (Skiba et al. 1989). Given Scotland's predominance of sandy and peaty soils, it is not surprising that most soils in upland Scotland in particular appear sensitive to acidification (Langan & Wilson 1993), and that the 'critical loads' for acidity (ie their ability to neutralise current acidic depositions) are exceeded (Department of the Environment 1992; Cm 2426 1994). Such acidification matters for several reasons. Soil acidification is generally reversible over time if the acidic inputs decline or stop, but the rates of change can be very slow (Wright & Hauhs 1991). Further, apart from the direct effects of acid pollutants on vegetation (Lee et al. 1989), decreases in soil acidity are likely to change the competitive ability of different plant species because of the increasing solubility of toxic aluminium and manganese, and of decreased rates of nitrogen mineralisation, so causing changes in vegetation composition, and with consequential further effects on topsoil acidity (Miles 1985). Although the indirect effects of acidic pollutants on vegetation via their influence on soils are imperfectly understood, they may currently be occurring across at least half of Scotland.

Study of the effects of acidic pollutants on soil acidity is complicated because many vegetation types themselves have strongly acidifying influences on topsoil (Miles 1985, 1986; Billett et al. 1988). Nevertheless, pollution-induced soil acidification in Sweden is estimated to cause a yearly loss in timber growth of 12 million cubic metres, with an associated loss of profit at current exchange rates of about £250-500 million (Sverdrup et al. 1993). I do not know of any comparable estimates for Scotland or for Britain as a whole.

Finally, despite growing evidence about the importance in recent decades of the effects of atmospheric acidic depositions in acidifying soils, it is still true that soils can become very acid under natural conditions because of leaching by rainwater and the acidifying effects of many types of herbaceous and woody vegetation, including woodlands (Johnson et al. 1991). The accumulated evidence that effects of different kinds of vegetation cover on the acidity and other properties of soils suggests that the reduction in Scotland's woodland cover in prehistoric and historic times from about two-thirds (though much of this cover would not have been dense) to perhaps as little as 1.1% today (Mackenzie 1987) and the replacement of much of this with more acidifying vegetation, in particular heather-dominated heathland and moorland, has acidified large areas of the Scottish uplands in particular (Miles 1985, 1988). No accurate estimates exist of the extent of this human induced soil acidification. However, in 1988 about 41 percent of Scotland had vegetation in which heather dominated or was abundant (Fig 12.1). Although some of this is peatland which has not been wooded for millenia, the figure excludes other upland areas, notably in the Southern Uplands, where former heather cover has given way to a predominance of grass. It therefore seems reasonable to suggest that about 30-40 percent of Scotland's soils, mainly in the uplands, are to some extent degraded because of acidification caused by past direct and indirect management. The main effects of this soil degradation appear to be reductions in biodiversity on local scales and in ecosystem productivity.

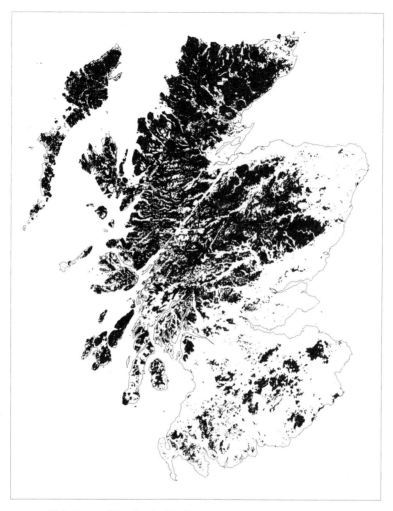

12.1: Areas of Scotland with dominant or abundant heather cover
(Land Cover of Scotland data)

Climate Change

There is a broad consensus that the world's climate may be gradually warming as a result of human activity increasing the concentration of greenhouse gases (i.e. gases whose molecules contain three or more atoms) in the atmosphere (Houghton et al. 1990, 1992). Current scenarios suggest that by the year 2050, the UK could be 3°C warmer than at present, and have perhaps 10 percent more precipitation, mainly in winter. The likely effects of these on the soil have been discussed in some detail (ITE/SSLRC 1989; United Kingdom Climate Change

Impacts Review Group 1991), and only some of the main points are mentioned here. Warmer summers without extra rainfall would lead to shrinkage of clay soils and an increased rate of cracking of buildings on them. However this seems unlikely to be a major problem in Scotland given the limited extent of clay soils. The most widespread effect on soils would be that warmer and wetter conditions would result in increased rates of organic matter decomposition and loss in soils. Lower soil organic matter contents would tend to result in: (1) lower water-holding capacities and hence soils more prone to drought; (2) lower soil stability and hence a greater risk of erosion and compaction; (3) increased rates of mineralisation of nitrogen, with possible implications for water quality; (4) poorer soil structure, and hence reduced rates of infiltration and increased run-off; (5) a reduced plant-nutrient storage capacity, especially in sandy soils; (6) reduced biodiversity in the soil; and finally (7) a feedback to global warming because of the conversion into carbon dioxide of carbon previously stored in the soil.

Conclusions

This necessarily brief and idiosyncratic *tour de table* of current problems with Scotland's soils suggests only localised problems in the lowlands but more widespread and chronic problems in the uplands. The lowland viewpoint may seem complacent, and may reflect a lesser knowledge of these soils; however it should not be taken to mean that there are no problems with soils in the lowlands of Scotland – this is certainly not the case. However, most of the problems are localised and are readily susceptible to control and amelioration by management. This contrasts with the uplands where widespread acidification has probably occurred as a direct or indirect result of past human activities and where sometimes profound ecological degradation can be inferred to have resulted. This is no new conclusion; McVean and Lockie (1969) argued this a quarter century ago, and the late Frank Fraser Darling a quarter century before that. There is certainly an argument that widespread soil and ecological degradation limits the land-use options of crofting communities and many upland estates today.

The good news about this past degradation is that, except for soil lost by erosion, most of the soil and ecological changes are probably reversible within 50-100 years or so if the vegetation cover is changed to types which tend to decrease soil acidity such as birchwood and bent-fescue grassland (Miles 1988). Indeed, compared with the soil conditions on many heather moorlands, even the much maligned Sitka spruce forest is a step towards reversing past decline because a single rotation will produce soil conditions more favourable to a wider variety of tree species than will now establish naturally on the open moor. Thus the roughly 13 percent of Scotland currently under plantation forestry, mainly with conifers, can be regarded as a significant step in the process of improving the quality of the soil resource. Further, Scottish Natural Heritage has made a start on trying to quantify the extent and severity of upland soil erosion and seems likely to take the matter further under its remit of sustaining the natural heritage.

Bibliography

Atlas R M. 1981. Microbial degradation of petroleum hydrocarbons: an environmental perspective. *Microbiological Reviews*, 45: 180-209

Atlas R M, Bartha R. 1992. Hydrocarbon biodegradation and oil spill bioremediation. *Advances in Microbial Ecology*, 12: 287-338

Bacon J. 1990. Atmospheric deposition of lead on Scottish hill and upland soils. *Macaulay Land Use Research Institute Annual Report for 1989-90.* p.42. Macaulay Land Use Research Institute, Aberdeen.

Bacon J. 1991. Atmospheric deposition of lead on Scottish hill and upland soils. *Macaulay Land Use Research Institute Annual Report for 1990-91.* p.32-3. Macaulay Land Use Research Institute, Aberdeen.

Ballentyne C K. 1991. Holocene geomorphic activity on the mountains of the Scottish Highlands. *Scottish Geographical Magazine.* 107: 84-98

Berrow M L, Webber J. 1972. Trace elements in sewage sludges. *Journal of the Science of Food and Agriculture.* 23: 93-100

Billet M F, Fitzpatrick E A, Cresser M S. 1988. Long-term changes in the acidity of forest soils in north-east Scotland. *Soil Use and Management*, 4: 102-7

Blumer M. 1976. Polycylic aromatic compounds in nature. *Scientific American* 234: 34-45

Blumer M, Blumer W, Reich T. 1977. Polycyclic aromatic hydrocarbons in soils in a mountain valley: correlation with highway traffic and cancer incidence. *Environmental Science and Technology*, 11: 1082-84

Bridges E M. 1988. Dealing with contaminated soils. *Soil Use and Management*, 3: 151-58

Burton R G O, Hodgson J M. (eds) 1987. Lowland peat in England and Wales. *Soil Survey Special Survey* No. 15

Cawse P A. 1980. Deposition of trace elements from the atmosphere in the UK. In: *Inorganic Pollution and Agriculture*, MAFF Reference Book 326. pp. 22-46. HMSO, London

Cheshire M, Shand C, Hepburn A. 1991. Radiocaesium pollution of soils. *Macaulay Land Use Research Institute Annual Report for 1990-91.* p. 33-4. Macaulay Land Use Research Institute, Aberdeen

Cm 2426. 1994. *Sustainable Development. The UK Strategy.* HMSO, London

Davies R D, Hucker G, L'Hermite P. 1983. *Environmental Effects of Organic and Inorganic Contaminants in Sewage Sludge.* Reidel, Dordrecht

Department of the Environment 1989. *Code of Practice for Agricultural Use of Sewage Sludge.* HMSO, London

Department of the Environment. 1992. *The UK Environment.* HMSO, London

Domsch K H. 1984. Effects of pesticides and heavy metals on biological processes in soil. *Plant and Soil*, 76: 367-78

Ecological Steering Group 1994. *The Environmental Impact of the Wreck of the 'Braer'* The Scottish Office, Edinburgh

Edwards N T. 1983. Polycyclic aromatic hydrocarbons (PAHs) in the terrestrial environment – a review. *Journal of Environmental Quality*, 12: 427-41

Environment White Paper 1990. *This Common Inheritance. Britain's Environmental Strategy*, Cm 1200. HMSO, London

Evans R, Bullock P, Davies D B. 1988. Monitoring soil erosion in England and Wales. In: Morgan R P C, Dickson R J (eds). *Agriculture: Erosion Assessment and Modelling.* pp. 73-91. Office of Official Publications of the European Communities, Luxembourg.

Forestry Commission 1993. *Forests and Water Guidelines.* 3rd edn. HMSO, London

Frost C A, Spiers R B. 1984. Water erosion of soils in south-east Scotland–a case study. *Research and Development in Agriculture,* 1: 145-52

Goodman G T, Pitcairn C E R, Gemmel R P. 1973. Ecological factors affecting growth on sites contaminated with heavy metals. In: Hutnik R J, Davis G (eds). *Ecology and Reclamation of Devastated Land,* Vol.2 : 149-71. Gordon & Breach, New York

Horrill A D, Livens F R, Beresford N A. 1989. Radionuclide transfer in terrestrial ecosystems. *Institute of Terrestrial Ecology Annual Report for 1988-89.* pp. 40-3. Natural Environment Research Council, Swindon

Houghton J T, Callander D A, Varney S K. (eds) 1992. *Climate Change 1992. The Supplementary Report to the IPCC Scientific Assessment.* Cambridge University Press, Cambridge

Houghton J T, Jenkins G J, Ephraums J J. (eds) .1990. *Climate Change. The IPCC Scientific Assessment.* Cambridge University Press, Cambridge

Hutchinson T C, Meena K M. 1987. *Lead, Mercury, Cadmium and Arsenic in the Environment.* Wiley, Chichester.

Innes J L. 1983. Lichenometric dating of debris flow deposits in the Scottish Highlands. *Earth Surface Processes and Landforms,* 8: 579-88

Institute of Terrestrial Ecology and Soil Survey and Land Research Centre. 1989. *An Assessment of the Principles of Soil Protection in the UK.* Vol. 2. *Review of Current Major Threats.* Institute of Terrestrial Ecology, Grange-over-Sands.

Interdepartmental Committee on the Redevelopment of Contaminated Land. 1987. *Guidance on the Assessment and Redevelopment of Contaminated Land.* Department of the Environment publication 59/83, 2nd edition.

International Agency for Research on Cancer. 1973. *IARC Monographs on the Evaluation of Carcinogenic Risk of Chemicals to Man 3: Certain Polycyclic Aromatic Hydrocarbons and Heterocyclic Compounds.* IARC, Lyons.

International Agency for Research on Cancer (1979). *IARC Monographs on the Evaluation of Carcinogenic Risk of Chemicals to Humans.* Supplement 1: *Chemicals and Industrial processes Associated with Cancer in Humans.* IARC Monographs Vols 1-20. IARC, Lyons

International Agency for Research on Cancer (1983). *IARC Monographs on the Evaluation of Carcinogenic Risk of Chemicals to Humans. Polynuclear Aromatic Hydrocarbons.* Part 1: *Chemical, Environmental and Experimental Data.* Vol. 32. IARS, Lyons

Johnson D W, Cresser M S, Nilsson S I, Turner J, Ulrich B, Brinkley D, Cole D W 1991. Soil changes in forest ecosystems: evidence for and probable causes. *Proceedings of the Royal Society of Edinburgh* 97B: 81-116

Jones K C, Stratford J A, Waterhouse K S, Vogt N B. 1989. Organic contaminants in Welsh soils: polynuclear aromatic hydrocarbons. *Environmental Science and Technology* 23: 540-50

Langan S J. 1993. The application of Skokloster critical land classes to the soils of Scotland. In: Hornung M, Skeffington, R A (eds). *Critical Loads: Concept and Applications.* pp. 40-47. HMSO, London

Leahy J G, Colwell R R. 1990. Microbial degradation of hydrocarbons in the environment. *Microbiological Reviews* 54: 305-15

Lee J A, Baddeley J A, Woodin S J. 1987. Effects of acidic deposition on semi-natural vegetation. In: *Acidification in Scotland.* pp. 94-102. Scottish Development Department. Edinburgh.

Macaulay Institute for Soil Research .1982. *1:250000 Soil and Land Capability Maps and Handbooks, Sheets 1-7.* Macaulay Institute for Soil Research, Aberdeen

Macaulay Land Use Research Institute. 1993. *The Land Cover of Scotland 1988: Ex-*

ecutive Summary. Macaulay Land Use Research Institute, Aberdeen

Mackenzie N. 1987. *The Native Woodlands of Scotland.* Friends of the Earth, Edinburgh

McCown B H, Brown J, Barsdate R J. 1972. Natural oil seeps at Cape Simpson, Alaska: Localised influences on terrestrial habitat. In: Institute of Arctic Biology. *The Impact of Oil Resource Development on Northern Plant Communities.* pp 86-90. Fairbanks, Alaska

McVean D N, Lockie J D. 1969. *Ecology and Land Use in Upland Scotland.* Edinburgh University Press, Edinburgh

Miles J. 1985. The pedogenic effects of different species and vegetation types and the implications of succession. *Journal of Soil Science* 36: 571-84

Miles J. 1986. What are the effects of trees on soils? In: Jenkins D (ed). *Trees and Wildlife in the Scottish Uplands.* pp. 55-62. Institute of Terrestrial Ecology, Huntingdon.

Miles J. 1988. Vegetation and soil change in the uplands. In: Usher M B, Thompson D B A (eds). *Ecological Change in the Uplands.* pp. 57-70. Blackwell Scientific Publications, Oxford

Miles J, Latter P M, Smith I R, Heal O W. 1988. Ecological effects of killing *Bacillus anthracis* on Gruinard Island with formaldehyde. *Reclamation and Revegetation Research* 6: 271-83

Ministry of Agriculture, Food and Fisheries. 1993a. *Code of Good Agricultural Practice for the Protection of Soil.* MAFF Publications, London

Ministry of Agriculture, Food and Fisheries. 1993b. *Review of the Rules for Sewage Sludge Application to Agricultural Land: Soil Fertility Aspects of Potentially Toxic Elements.* Independent Scientific Committee. MAFF Publications, London

Moffat A J. 1988. Forestry and soil erosion in Britain–a review. *Soil Use and Management* 4: 41-4

Morgan R P C. 1987. Sensitivity of European soils to ultimate physical degradation. In: Barth H, L'Hermite P (eds). *Scientific Basis for Soil Protection in the European Community.* pp. 147-60. Elsevier, London

Moriarty F. 1988. *Ecotoxicology. The Study of Pollutants in Ecosystems.* 2nd edn. Academic Press, London

Nature Conservancy Council. 1990. *The Impact of Afforestation and Forestry Practice on Freshwater Habitats.* Nature Conservancy Council, Peterborough

Paterson E. 1989. Soil Protection. *Macaulay Land Use Research Institute Annual Report for 1988-89.* pp 15-23. Macaulay Land Use Research Institute, Aberdeen

Proctor J. 1971. The plant ecology of serpentine. III. The influence of a high magnesium/calcium ratio and high nickel and chromium levels in some British and Swedish serpentine soils. *Journal of Ecology* 59: 827-42

Purves D. 1985. *Trace Element Contamination of the Environment.* Elsevier,Amsterdam

Royal Commission on Environmental Pollution. 1983. *Lead in the Environment.* Ninth report. HMSO, London

Salmon Advisory Committee. 1991. *Factors Affecting Natural Smolt Production.* Ministry of Agriculture, Food and Fisheries; Scottish Office Agriculture and Fisheries Department; and Welsh Office Agriculture Department. London

Scott V J R. 1993. *A study of soil erosion on the Trotternish Ridge of the Isle of Skye.* MSc thesis, University of Aberdeen

Sims R C, Overcash M R. 1983. Fate of polynuclear aromatic compounds (PNAs) in soil-plant systems. *Residue Reviews.* 88: 1-68

Skiba U, Cresser M S, Derwent R G, Futty D W. 1989. Peat acidification in Scotland. *Nature* 337: 68-9

Smith M A. 1988. Reclamation and treatment of contaminated land. In: Cairns J (ed).

Rehabilitating Damaged Ecosystems: Vol.1. pp 61-89. CRC Press, Baton Rouge

Soutar R G. 1989. Afforestation and sediment yields in British freshwaters. *Soil Use and Management* 5: 82-6

Spiers R B, Frost C A. 1985. The increasing incidence of accelerated soil water erosion on arable land in the east of Scotland. *Research and Development in Agriculture* 2: 161-67

Scottish Office. 1991. *The Scottish Environment: Statistics.* No.3. The Scottish Office, Edinburgh

Sverdrup H, Warfvinge P, Jonsson C. 1993. Critical loads of acidity for forest soils, groundwater and first-order streams in Sweden. In: Hornung M, Skeffington R A (eds). *Critical Loads: Concept and Applications.* pp 54-67. HMSO, London

Symon J A. 1959. *Scottish Farming, Past and Present.* Oliver & Boyd, Edinburgh

Thornes J B. (ed). 1990. *Vegetation and Erosion Processes and Environments.* Wiley, Chichester

Thornton I. (ed). 1983. *Applied Environmental Geochemistry.* Academic Press, London

United Kingdom Climate Change Impacts Review Group. 1991. *The Potential Effects of Climate Change in the United Kingdom.* HMSO, London

Van Ness G B. 1971. Ecology of anthrax. *Science* 172: 1303-7

Vergnano O, Hunter J G. 1952. Nickel and cobalt toxicities in oat plants. *Annals of Botany.* N.S. 17: 317-28

Wild S R, Berrow M L, Jones K C. 1991. The persistence of polynuclear aromatic hydrocarbons (PAHS) in sewage sludge amended agricultural soils. *Environmental Pollution* 72: 141-57

Wilson S C, Jones K C. 1993. Bioremediation of soil contaminated with polynuclear aromatic hydrocarbons (PAHs): a reveiw. *Environmental Pollution* 81: 229-49

Wolf K, Van den Brink W K, Colon F J (eds). 1988. *Contaminated Soil '88.* Kluwer, Dordrecht

Wright R F, Hauhs M. 1991. Reversibility of acidification: soils and surface waters. *Proceedings of the Royal Society of Edinburgh* 97B: 169-91

INDEX